THE BOOK OF GRAPHIC PROBLEM-SOLVING

THE BOWKER GRAPHICS LIBRARY

THE BOOK OF GRAPHIC PROBLEM-SOLVING

How to get visual ideas when you need them

JOHN NEWCOMB

R. R. Bowker Company
New York and London, 1984

For Charlotte

Published by R. R. Bowker Company
205 East Forty-second Street, New York, NY 10017
Copyright © 1984 by Xerox Corporation
All rights reserved
Printed and bound in the United States of America

Library of Congress Cataloging in Publication Data

Newcomb, John, 1937–
    The book of graphic problem-solving.

    Includes bibliographical references.
    1. Printing, Practical—Make-up.   2. Printing, Practi-
cal—Layout.   3. Magazine design.   4. Graphic arts.
I. Title.
Z246.N555   1984   686.2'24   84-11066
ISBN 0-8352-1895-3

CONTENTS

PREFACE

New things from old
The light bulb, ageless symbol of inspiration, is still alive and well in modern design; the uses on the next few pages will testify to this. The photo on the facing page was created for self-promotion for the photographer. The image could be called "the respectable idea."

There were two main reasons for writing this book. The first grew from a need for self-defense: For the last three years I have been the design director of Medical Economics, a company that publishes several magazines and books for doctors, nurses, and other health care professionals. Most of these magazines are published monthly, but a couple—including the flagship book with the company name—are issued biweekly. *Medical Economics* magazine, a general financial and marketing guide for physicians, is published every two weeks of the year and, due to an unprecedented surge in pharmaceutical advertising, has been running in the maximum range of 300 to 350 pages each time for most of the last three years. The magazine features a concept or analogic cover—an idea cover—each time, and many of the articles utilize conceptual art as well. Further, several of the other publications in the group require interpretive art on the cover or for inside features.

What quickly became apparent was that my staff of designers and I had to set up a true idea mill to find reliable ways to produce graphic ideas in bulk and on schedule. By conservative estimate, we had to find dramatic solutions for 125 covers and 350 inside visuals each calendar year. Some sort of conceptual method was needed if the magazines were to flourish and if the staff was to keep up this torrid pace of creativity. We needed to codify an approach to each visual problem that would yield high-quality solutions for stories dealing with subjects that would not differ in major ways from year to year. Without such a method my staff of talented art directors and I would be facing a problem of burnout that could reach epidemic proportions.

So I devised something to fill the need. This approach, which I call the Bite System, is a largely mechanical, word-oriented way of analyzing a central idea and sorting out the various implications in an author's statement, the working title. This method will be described and demonstrated later in the book.

My second reason for writing the book was a series of lectures I was asked to give for *Folio* magazine's Face-to-Face publishing seminar series. This week-long program of talks is given each fall in New York; the speakers are top specialists in all different areas of publishing—designers, editors, writers, and production and sales people—who share their knowledge with the enrollees.

Jan White, a design consultant and a good friend of mine who is active in the program, suggested to the *Folio* executives that I might give a talk on magazine art directing, so the conference manager and I discussed possible subjects. As I scanned the conference program of the previous year, I noticed that the fundamentals of editorial design had been thoroughly covered: There were courses on most of the technical aspects of art directing—specifying type, cropping photographs, using grid·patterns, understanding the limitations of press imposition—as well as ones that dealt with redesigning magazines, creating effective covers, arranging distinctive contents pages, departments, or features. There were even lectures on art directing for nonartists!

I was struck by the fact that all of the design talks seemed to deal with the physical form of editorial design, the style of the page, the choice and handling of illustration, photography, and type. The glaring omission was in the most crucial area of art directing: getting ideas for exciting and effective graphic presentations. The courses that discussed the surface elements of designing, valuable as they were, presented information of the kind that any middle-echelon graphics professional had mastered long since. None of them came close enough to the real magic of the trade.

This menu of lectures had nothing about ways to build ideas, nothing about how to use graphics to provoke curiosity, to entertain, to engage the reader's self-interest and reinforce the writer's message—all

the parts of a designer's job that make it satisfying and that are rewarded by employer and peers alike. Once the ideas exist, executing them—picking and directing artists and photographers—is simply a choice of weapons, largely a matter of careful follow-through and control of detail. Far more vital is the creation of visual elements that reinforce the verbal message, the making of images that give the writer's words the force needed to lodge them in the reader's mind.

There is no skill more important to an art director than this one of getting ideas for graphics; this is the special ability that gains the respect of the business world. A designer who doesn't have this skill becomes little more than a layout machine, a page decorator, a graphics faucet to be turned on when the need to package is felt. Such a person may be extremely facile at processing what an editor hands him, but much, much more is needed to make effective editorial statements that are remembered once the magazine is put down. The artist must train to become a full partner in the business of communication. *What* is finally said, as much as *how* it is said, must be the designer's concern, too. George Lois, the advertising art director, puts it well

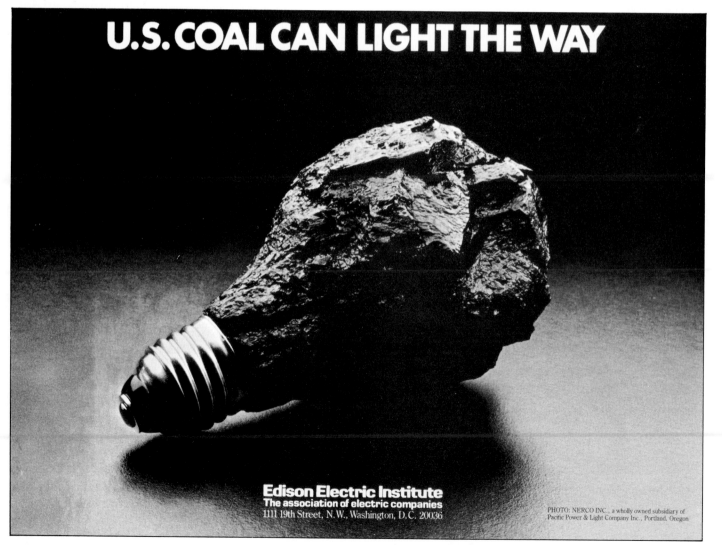

The alternate-energy idea.

when he says, "All the tools in the world are meaningless without an essential idea. An artist without an idea is unarmed."*

So the need was there. I suggested to the *Folio* program manager, Rob Sgarlata, that I might do a talk on methods of getting ideas, on the flashier parts of an art director's job. He was excited but expressed some doubts: could I really give advice that would be useful to people working in a field marked by near total diversity? Professional art jobs differ wildly from place to place; a New York agency art director would seem to have little in common with the art editor of a Midwest trade magazine or with a promotion designer huddled in a cramped corner of a general office building in downtown Philadelphia. Could a core process of creativity be distilled and described that would have meaning for all of these graphics people? This had been the main obstacle for giving such theory lectures in the series before; attempting a project of this type would require considerable courage. It would be easy to look foolish, to

*George Lois and Bill Pitts, *The Art of Advertising* (New York: Harry N. Abrams, 1977), p. 324.

Parish Kohanim

The high riding idea.

appear to be little more than a quack, a peddler of graphic cure-alls.

But the program manager stifled his doubts and I plunged into the homework. I had my own print design experience on various editorial teams from which to draw—twenty-odd years of creating books, magazines, and advertisements for some of the best publishers in the world—but I felt that talking shop with fellow designers would furnish new technical insight for the lecture. Soon I was speaking with some of the most famous people in editorial and advertising design. Then the subject took on a life of its own, and other kinds of professional idea-getters were giving their enthused response: illustrators, syndicated newspaper cartoonists, even copywriters wanted to discuss the process of getting ideas. They wanted to talk about problems they all faced and to join in the hunt for common factors in creativity.

Gradually, some patterns began to emerge as well as an abundance of useful material for the seminar. The talk, which I called "The Art of Art Directing," was a popular one. Fall of 1983 was the third time I gave it, and each year new information and new insights broadened its scope. To shorten the story, the subject cried out to be put in book form, to become a permanent part of the professional literature.

I owe a great deal to many people because of this book. My friend Jan White, a fine author and designer himself, not only got me involved with the *Folio* talks but also made the connection with the publisher of his own books, *Editing by Design*, *Designing for Magazines*, and *Mastering Graphics*. He has a habit of bringing turmoil—and growth—into my life every few years.

To my editor, Betty Sun, and the efficient, supportive people at R. R. Bowker go my deepest thanks. I also wish to thank Gretchen Bruno, my design assistant on this book; she made a major and valuable contribution to the project.

I'm grateful to Medical Economics Company for allowing me to work on the book and for letting me use such a large amount of work done for the company magazines to demonstrate the theories. I must also express my thanks to the art directors on my staff, whose working lives became living laboratories for design theses.

Finally, I'm deeply indebted to the many top creative people who gave so freely of their time and furnished examples of their best work. All of them were forthright and unselfish when it came to sharing their professional insight. The book wouldn't have been possible if they hadn't talked freely about the little quiet things that happen when a designer and a problem come face to face.

INTRODUCTION

Anything we don't quite understand arouses awe and wonderment in us: think how impressive the plumber's skills seem after that infuriating drip-drip of the faucet has been stifled at last. Don't we call it "magic" when the mechanic lifts the hood, tinkers with a bunch of wires, and the recalcitrant motor springs to life? That's just the reaction I had when watching Newcomb at work. He may not fix faucets—though for all I know he may understand how motors work—but he does know how to think up graphic ideas.

I first met Newcomb when attending editorial planning meetings as a consultant for the magazine on which he worked. These conferences with the editors typically began as sackcloth-and-ashes sessions; 20/20 hindsight was used to figure out ways that stories could have been improved in their visual/verbal presentation. Then the agenda would turn to upcoming articles, and that's when he would start bubbling ideas—some wild, some silly, others that were familiar images with unfamiliar twists—all to the editorial point and technically feasible. I was awed by the performance because my mind is oriented less toward that aspect of publication design (the creation of graphic ideas) than toward

BATESWORLD FALL 1981

idé

The homespun idea.

The emerging idea.

Bjorn Winsnes

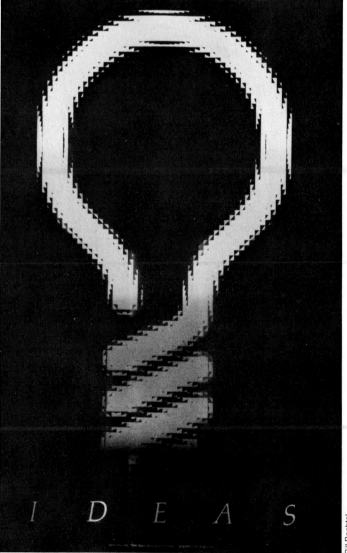

I D E A S

Bill Bechtel

x

the format into which they are set. Newcomb agrees that the context is important for successful presentation; after all, you want to make those good ideas as visible as possible. However, he derives more satisfaction from the jewels (the individual images made with illustration and photography) than from the setting (the publication's typography and editorial grid).

What sort of an animal is this, I asked myself, and what makes a mind work this way? How can such a stream of visual combinations and graphic ideas be turned on and off apparently on demand? As I watched and got to know him better, I could see that several characteristics (none of which is especially rare or peculiar, but whose combination is unusual) were at work:

1. Intellectual breadth. Since he is a voracious reader, his bank of knowledge furnishes raw material for building visual combinations. He laughs about this, saying that his head is so full of random facts that if it could be opened, trivia would spill out much like junk from Fibber Magee's closet. Random facts they may well be, but I see this broad-based human involvement as a major asset essential to his success as a designer.

Bruno R. Schreck

2. Capacity to concentrate. This lets him strip away the superficial and home in on the essence as he approaches solutions. That, in turn, allows him the luxury of churning up many choices in the beginning of the creative process, when he can be less critical of the ideas; then, switching mental gears, he can be ruthless in editing out unusable thoughts.

3. Ability to communicate visual ideas. He draws. He was trained as an illustrator and he can sketch an idea almost as fast as he can describe it. His explanations wander from words to sketches on yellow pads—he shows-and-tells. Furthermore, he is a skilled photographer confident of his judgment in lighting, staging, and composition. As a result, the visual solutions he invents are invariably practical: he understands the physical possibilities of a concept.

4. Love of language. He moves easily in the world of words and his respect for written and spoken thought gives vitality and luster to his graphic solutions. He knows that words are the foundation on which pictures, diagrams, conceptual art—in fact, all communication visuals—are built. Verbal language defines the purpose, direction, and meaning of the message embodied in the visual symbolism the designer produces. I suspect that this is where our minds meet: we are both convinced that the more *verbally oriented* you are as a designer, the better your images will communicate. By the same token, we know that visual responsibility on the part of verbally oriented editors is equally essential: the more *visually oriented* you are as a writer/editor, the more cleverly you can guide and help the image-maker. A high regard for words is the ingredient that helps Newcomb penetrate to the core of a problem.

So I watched him perform these feats of mental magic and the nature of the ability wasn't revealed until he wrote this book. Like that auto mechanic or plumber or any good professional revelling in his craft, he naturally claims that it's easy doing what he does. It's just a matter of knowing how. That is perhaps a trifle optimistic. But this book is an excellent guide to his system. He has defined the steps to his approach, making it accessible to us who are awed by such expertise. I've tried the method. It works.

Jan V. White
Westport, Connecticut

The mercenary idea.

Elwood Smith was asked to make a drawing for the
Push Pin Graphic *magazine on the theme of getting
ideas. He wrestled with the problem until it became
clear that the illustration should really show the
opposite of the theme: the designer who* can't *get ideas.
The world swarms with inspiration, but our man stays
locked in his own blue funk.*

WHY IS THIS BOOK NEEDED?

We've all been there. The art conference starts in ten minutes and you *must* have an idea for the next cover. In fact, you need two or three ideas for the cover. Nothing comes. Memories of last night's televised football game muscle into your mind. You mutter a curse and yank your thoughts back from Howard Cosell. An idea . . . *any* idea. "What can we do?" you ask yourself for the tenth time in five minutes. You stare at the layout pad in front of you. It's blank. You look at the drawing board in front of you—blank. You shift your gaze to the wall in front of you—blank, blank, blank!

We've all been in this spot often enough. It's never pleasant to be stymied as you approach a deadline; it makes your palms perspire and does funny things to your stomach. But this agony can usually be avoided, and this book is dedicated to that purpose.

Getting good, usable graphic ideas is the most important job a designer has, and the ability to perform consistently in this area is the largest factor in his or her professional success. This skill often seems magical: to the average person, the process of creativity seems like a surge of electricity, a flash of heaven-sent fire that works miracles at the behest of a special and exotic creature, the designer. This line of work is heavily wrapped in myth and mystique, and top practitioners are seen as wizards. Designing seems more like something semidivine that happens to the lucky few than something to which mere mortals may aspire.

It is essential to dispel such notions from the very outset, for getting ideas is a craft, not an art. This craft, like any other, can be taught and learned; it has method and an accepted, tested logic. Like anything else that's learned—say, playing the flute or riding a horse—it requires study and practice, but it is accessible to anyone of average intelligence who will invest the time and effort needed to sharpen this skill.

All of the leading graphics professionals interviewed for this book had mastered this craft and developed their own ways of using it, but each of them had groped his or her way to competency; each had to search by trial and error to reach the highest levels of skill, and each had to search alone. There has been little logic developed or research done on how a designer can go about purposely developing the ability that makes him or her a sought-after asset in the commercial world. Most art training has been case-oriented; that is, famous designers present their solutions to specific problems. The "I-was-there-and-this-is-how-I-handled-it" method of teaching has very limited value for the beginner. Unless the student eventually faces exactly the same kind of client, has the same message to convey, and operates with the same restrictions or possibilities as in the case example, there is not much substantial guidance to be gained from this approach.

Many of these leading graphic artists seemed mildly

puzzled as to how they got where they are. Their career explorations were solitary, random, almost hit-or-miss in character. They developed instinctively, intuitively, without much awareness of the ingredients of their success. All of them were bright enough, lucky enough, and practical enough to make the most of whatever jobs they got; each one, with his or her own creative method, found a commercial home. When they were asked to describe their methods of solving visual problems some of their first remarks were revealing: they said, "Some days I have it, some days I don't"; or "I don't know. The magic is usually there when I need it"; or "I'm just crazy and I'm lucky I can use it in my work."

But as we talked, it became apparent that there was consistency in the experience of these successful artists and designers, and patterns of work method, training, and conditioning started to emerge. These patterns are understandable, nonmagical progressions of common sense. The methods are within reach, and conditioning can be learned. There are ways of getting good, viable graphic ideas, and what these practitioners of the visual arts had to say about their working environment, their mental preparation, and their idea-getting mechanisms gave substance to a fairly universal design approach.

Equally important, what they said gave validity to a couple of simple techniques I had devised to help the designer overcome the artist's equivalent of writer's block, the dread blank pad syndrome described at the beginning of this chapter. And these techniques will be useful to the veteran designer as well as to the beginner. The systems do two things: first, they outline a specific and practical method for attacking a given graphic problem by prescribing ways to define a message and sort out its usable parts; second, they improve the designer's associative and synthetic capacity, his or her ability to mix attributes and create unexpected images. They are designed to build the brain in special ways.

The intellectual conditioning described in this book can be of value to anyone who plans a creative career, anyone who wants to make a living originating and conveying ideas. Editors, copywriters, playwrights, novelists—all such professionals can benefit from it even though it is aimed primarily at those responsible for picture making—art directors, designers, and illustrators. The process of generating ideas is much the same for both writers and visual artists; in fact, since the roles of both overlap and since often both are working together to create the final piece of communication, it makes excellent sense for each to study the same techniques of idea production.

It is important for graphics specialists and professional writers to work together from the very beginning to create effective communication. The image-makers must be given access to the word turf; they must be allowed to massage headlines, offer better adjectives, suggest alternate expressions. Writers and editors, on the other hand, must not meet contempt from artists when they venture to express their ideas in doodle form or when they suggest a change of visual detail in an illustration or photograph. The creative team must trade ideas freely and, at least in the planning stages, ignore their specialties. The creative planning session is a meeting ground for intellectual partners, and the only measure of quality must be the ideas that emerge. The goal of all participants is effective communication, the best way to reach and influence the target audience.

What makes good communication? The magazine spread, television commercial, or print ad that is remembered longest after exposure is one in which words and pictures are applied to the same purpose. It is one in which the picture "reads" as easily as the words, where both elements—words and pictures—reinforce the same message. Most often the headline of a story should function as a colorful caption to the first illustration or photograph. Ideally, the opening pages of an article or an ad page will deliver a verbal *and* visual one-two punch that gives far greater impact than either words or pictures alone could produce. The double-barreled impact of a well-harmonized headline and visual element is the best way to make the message stick in the reader's mind long after he or she has put down the magazine or switched off the television set.

George Lois, the great advertising man, makes this observation: "An art director must be someone who treats words with the same reverence that he accords graphics, because the verbal and visual elements of modern communication are as indivisible as words and music in song."* Pictures and words together: this is the advice given by a famous picture-maker.

Jan White calls the visual/verbal character of good graphics the "yin-and-yang" relationship of words and pictures. He says that two opposite languages are involved, but that anyone who neglects to use both to maximize the message misses a chance to make one and one equal three. If this is true, though, why is the communications landscape so crowded with uncoordinated graphics—with magazine spreads that confuse and mislead, with inane ads that are instantly forgotten? Clearly the world is littered with missed chances. By contrast, the ads, covers, and pages in this chapter have been chosen to demonstrate how much thought and care can be given to making words *and* pictures carry the idea.

It may come as something of a surprise to hear that graphic problem solving should be approached on the word side rather than the picture side, that you should begin with words rather than images. An appreciation for words, the music of written language with its subtle range of shades and tints, can and should be cultivated early on even by those who are drawn to images. One way to develop this sensitivity to words is a regular regimen of reading.

Effective communication requires both the right words and the right pictures to attain maximum power; the two *must* work together. What I hadn't realized until I started seriously trading information with other designers is the importance of words at the beginning of the creative process.

The average person may think of graphic design as a picture-first, inspirational ritual and the designer as one who, by virtue of some specialness (read: eccentricity), has wonderful images float into his or her head from time to time. Nothing could be further from the truth. Top designers and idea-getters are disciplined professionals who have found their craft—and the foundation of their craft is words.

Mary K. Baumann, the art director of *Geo* magazine, puts it this way: "A lot of designers don't get along very well because they can't think in terms of words and also, on the other side of the coin, there are a lot of editors who can't think in terms of pictures. Really, the perfect marriage is between people who can think about both of them in terms of their function. If you can start that process back when the story starts to happen, it's even more exciting and that's where it's got to start. You've got to deal with the information you want to impart to your reader."*

Early verbal *and* visual planning, which you begin by conceptualizing with words, helps to increase the yield of graphic ideas and keep the graphic statement on target. Logically, the planning process starts with words: let's assume your editor or author has an idea for an article. This idea—like most thoughts—is made of words, carefully chosen words that are the bricks with which the writer builds a message for the reader. The main point of this message will probably be embodied in a working title for the story. The words the writer chooses will normally carry enough specifics, enough shades of meaning, and enough action to get the point across quickly.

It is from this statement, or perhaps from some expression in the story that makes the point in a different way (and could alter the headline later if the message is not changed) that we start making a graphic solution to enhance the writer's statement. We do *not* start with a free-floating, striking picture idea that so

---

*George Lois and Bill Pitts, *The Art of Advertising* (New York: Harry N. Abrams, 1977): 324.

*Jim Nelson Black, "Magazine Design: The Evolution," *Folio* (November 1983): 87.

enamors us that we forcibly graft the image onto the author's thought. This is a low-yield technique that often leads to shotgun marriages with words that don't fit. Such ill-matched pairs usually dilute the statement on the page. At best the pictures and the headline form a non sequitur—they have no apparent relation to each other. Sometimes the pictures and words pull in opposite directions and the poor reader doesn't get any message at all; he simply turns the page.

The process of problem solving begins when the editor drops a manuscript on your desk, gives you one or two working titles as well as a short explanation of the story, and asks you when an art conference should be scheduled to plan the layouts. It may very well be that perfectly good documentary illustration is available and dramatizing the feature will be no problem at all. Perhaps a first-rate reportage photographer has been assigned to take pictures of the people and places in the story; if so, the layout is simple. But perhaps all you're given are some hazy Polaroid prints taken by the author.

Or perhaps the story has to explain a theory. In these cases the designer and editor must exert themselves to create eye-catching graphics.

A word of caution is in order here. If, after reading the article, you're confused about the purpose of the piece, by all means talk it out thoroughly with the editor and the author. Take as much time as you need to discuss the story until you and your editor have a clear—and similar—idea of the communication job to be done. Many a brilliant graphic solution has been knocked off course by failure to understand at the very beginning of planning what it is, exactly, the editor wants to say to the reader.

But let's assume you've done this and that you're ready to start mining the words for graphic gold. You go off to a private place to do your homework so that you and the editor will be able to trade your best ideas in the art conference. Now you've reached the critical stage in the process of building picture ideas. Let's get down to business—creative business.

WORDS AND PICTURES, PICTURES THAT READ

On the next few pages are examples of print design that make effective use of both words and pictures to carry their messages. What each of them has to say is strong, unmistakable and vivid. Each gives ample evidence of the care taken to harmonize the visual and verbal elements to make a clear point, although in most cases one look at the picture is enough to catch the idea. Note also the frequent use of humor to lend entertainment value to the messages.

Two of a series of illustrations for Adidas sport shoes show the effective use of image mixture. On this page the message "Do your running in New York" is dramatized by the foot of a jogging giant landing in Central Park; the foot is clad in the New York model Adidas running shoe. On the opposite page the Adidas tennis shoe is linked to the terms cushion *and* comfort *because the tennis court is portrayed as a huge mattress. (This is an appealing idea for avid players, who often wish their feet would last for just one more set.)*

Wherever you live, run in New York.

adidas reports:

Whether you live in Chicago, Los Angeles, in Montana or Maine, you ought to run in New York —our new New York shoe—because it has features that actually lessen your chance of injury. Like an extended heel counter and solid carbon rubber outsole with a concave profile. Both help lead your foot and insure proper follow-through. This is important because proper follow-through reduces pronation and stress. The New York is exceptionally comfortable, too, thanks to our unique adi-air midsole. What's more, we've even placed an EVA heel stabilizer at the heel strike position for the rear foot control every runner needs. Not only does this shoe provide great control and protection, but it's just about the lightest, most flexible training shoe around. So no matter where you live, run in New York.

EVA heel stabilizer.

The New York

adidas

Wayne McLoughlin

4

Willardson and White Studios

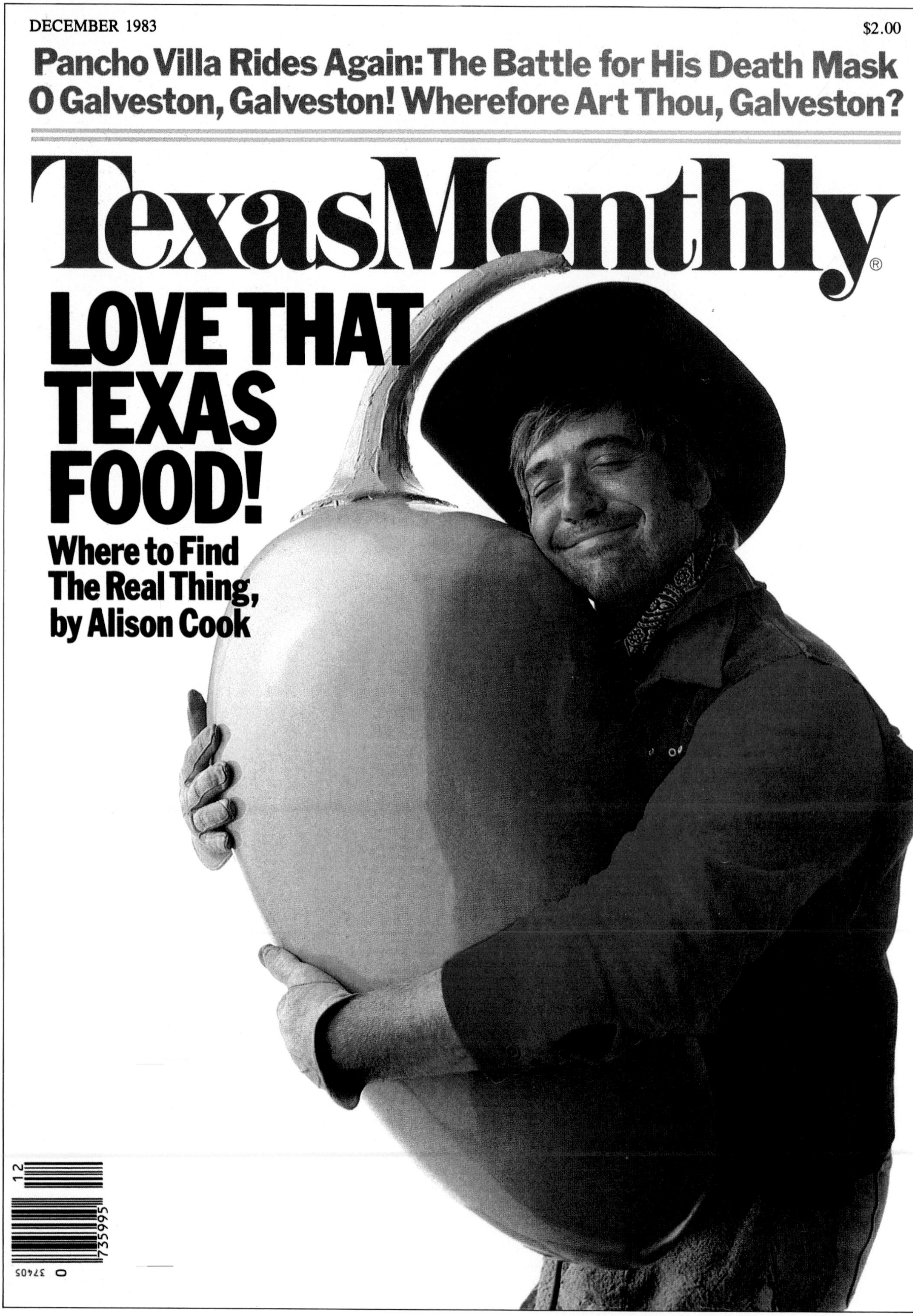

Robert Latorre (art by Joseph Melancon)

The Jalapeña pepper is the most distinctive ingredient in Tex-Mex cuisine, so the designer made it the focal element. This cover is guaranteed to make any native Texan grin.

Doctors burn out as often as anyone else whose job subjects him or her to relentless pressure. In this case, the statement was given uniqueness by a simple trick: replacing the lens of ordinary eyeglasses with lemon slices. The headline was adjusted to use the expression "turns sour."

WHAT TO DO WHEN MEDICINE TURNS SOUR

The outlook turns grim at some point in most medical careers, our survey shows. Here's how your colleagues regain their focus.

By Stephen Kaufman SENIOR ASSOCIATE EDITOR

"I've grown to hate medicine. I just can't wait for the day I retire."

"If I had anything else to fall back on, I'd leave my practice. I'd rather own a small restaurant, or repair cars, or do just about anything else."

"If I had it to do over again, I'd never invest all that time and effort in medicine. It isn't worth it."

Whether you call it mid-life crisis, or burnout, or just the blahs, feelings of dissatisfaction with medical practice aren't uncommon among private physicians. If you've experienced such despondency, take comfort from responses to a MEDICAL ECONOMICS spot check—you're not alone: 67 percent of your colleagues said, "Yes," they've felt disenchantment with medicine at some point in their careers (and, one respondent said, the other 33 percent were probably lying). Take comfort, too, that you'll probably come out of the doldrums, like many of your colleagues, as a better doctor and a happier person.

How? The most successful were those able to identify the problem—to zero in on the vague feelings of discontentment—and then prescribe a solution. If you ever start to feel that being a doctor is unfulfilling, their experiences may help you recover your focus. Here's how some of our survey respondents bounced back:

Take a new look at what you can contribute

The Oregon ophthalmologist stood at the door of the medical-mission building in Haiti, wondering whether the intense heat and hard work were worth giving up three weeks of vacation. Down the road, a mother led a stumbling boy toward the mission. Another poor kid blind from malnutrition, he thought.

"But when I examined him," the doctor recalls, "I found he was merely nearsighted—just like me." The mission didn't have any more glasses, so the doctor pulled out his own spare pair.

"The boy put on my gold-framed glasses, and looked around. He could see, maybe for the first time in his life. I'll never forget the expression on his face."

Answering an ad for the mission project had been a desperation move for the ophthalmologist, who'd become fed up with the daily routine of his practice after 12 years and was seeking "something more." At first, the hopelessness in Haiti—children dying from lack of medical care, parents expecting miracles from the foreign doctors—had plunged him into a deeper despondency.

"I used to ask myself why I did it. I was miserable in Oregon, miserable in Haiti. But that kid with my glasses did as much for me as I did for him. Work in Haiti took on a purpose, and when I got back to Oregon I was a different person. I seldom get irritated with patients' petty complaints anymore. When I do, I think about that kid walking the mountains of Haiti with my gold-framed glasses. Hell, they must represent a year's income around there. The thought makes life worthwhile."

If you can't beat them, don't join them

When the Denver physician finally sobered up, he decided he'd bet-

162 MEDICAL ECONOMICS/OCT. 11, 1982

S H A T T E R E D D R E A M S

Baccarat

These ads for Baccarat glassware ran in magazines targeted to upscale audiences, people who know the value and quality of Baccarat products. Rather than show the expected image of dreamily lit, lovingly photographed glass vases sitting in stately security in a fine home, the designers chose to show impending disaster.

S H A T T E R E D D R E A M S

Baccarat

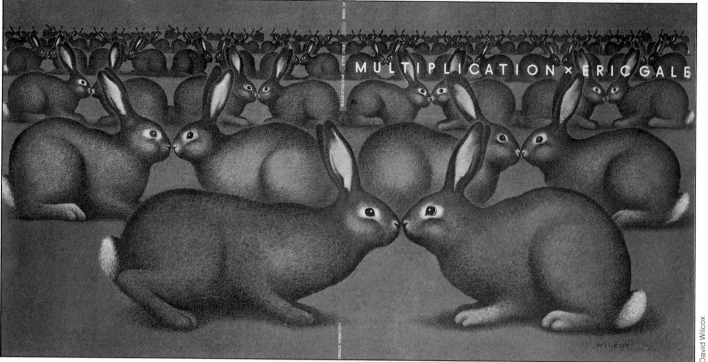

David Wilcox

Each of these three jazz album covers was designed by Paula Scher. In each case the name of the album served as a springboard for visual statements delivered with wit: multiplying rabbits, burning toast, and food for the ear.

Richard Hess

David Wilcox

8

Robert Krogel

This ad for a leading insecticide is engaging mainly because it uses the old maxim "hear no evil, see no evil, speak no evil" to humorous advantage. The pun works well and the body language of the multi-legged weevils adds an extra comic kick.

Robert Grossman

Anyone who's ever made a long trip in an airplane knows firsthand about jet lag, the nagging fatigue that results when your body's internal clock is out of sync with the time zone you are in. Quite literally, some parts of your body are moving faster than others when you leave the plane.

9

Ben Somoroff

A classic cover shot by the late Ben Somoroff brings each viewer to a place he or she has known well — the moment when the after-taste that usually follows a sip of artificially sweetened cola sets in. The casting and body language of the model are perfect.

Marvin Mattelson

Love is rather easily converted to hate. The artist shows the process of conversion with a red heart that melts and drips to form a poison-green, heart-shaped puddle on the floor.

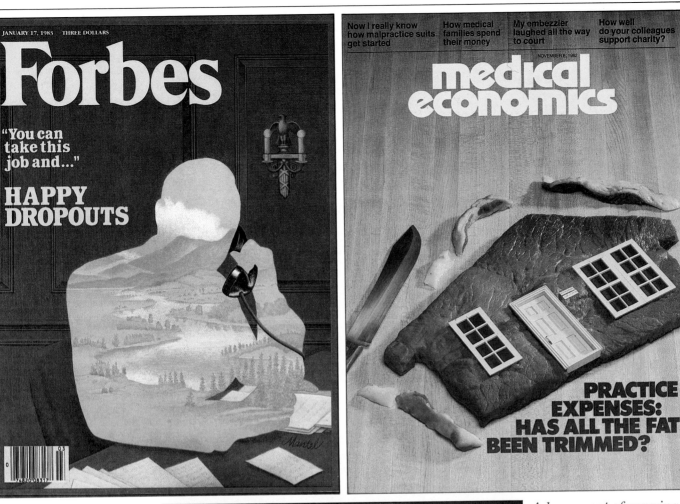

The executive suddenly decides that what he's doing every day is not worth the effort, that his time could and should be better used. The artist shows the man as a vacant spot, a window.

A large part of running any business is tailoring overhead to prevent unnecessary spending, while allowing enough money for operation. This cover asks the doctor whether he's looked closely enough at his routine costs.

A cat without Kitty Litter® Brand seems like a whole different animal.

It's not kitty's fault if kitty's cat box doesn't smell nice and fresh. It's the ordinary cat box filler you're using.

What you both need is Kitty Litter Brand. Because it's moisture activated.

When kitty does what kitties do, a special deodorant is released. A deodorant that eliminates every trace of odor. No other brand has it.

There's only one kitty in the world like yours. Buy him the one cat box filler that totally eliminates odor. Kitty Litter Brand.

There's only one Kitty Litter® Brand.

Is your cat a stinker? This quick shot to the pet owner's head says that without Kitty Litter products there is such a risk. Elegant photography and gentle wit make this ad stick in the mind.

Mark Kozlowski

Your floor is safe, says this ad for a leading brand of durable floor covering. Even if the best laid plans of mice and little hostesses of birthday parties go astray, the damage is limited. The clever staging of the set helps grab our attention; the model is seated and several of the flying props are suspended by sticks or wires.

World economic outlook: Commodities and the LDCs Page 126

A McGraw-Hill Publication

BusinessWeek

November 9, 1981 • $1.75

WHY DETROIT STILL CAN'T GET GOING

Page 106

Braldt Bralds

When this cover was done, American carmakers were wallowing in a seemingly hopeless morass of marketing trouble. Imported cars were pummeling them with evidence of better workmanship and more consumer-aware design. There were skeptics who said that Detroit was stalled permanently. Braldt Bralds's all-neutral gearshift made the point well.

13

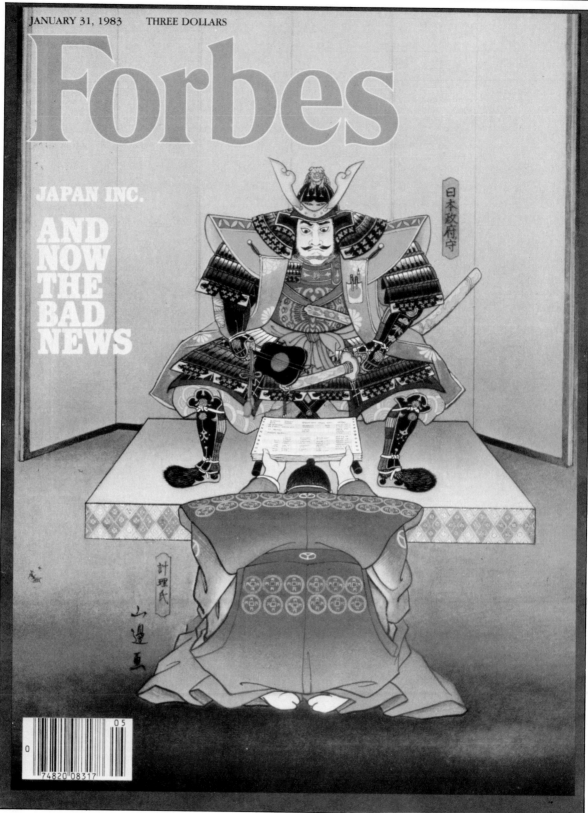

This is a skillful blending of ancient and modern. A scowling warlord representing Japanese business is glaring at an obsequious servant who's handing him the latest computer printout for "Japan, Inc." The ink on the report is red.

INCOME TAXES

IS YOUR TAX ADVISER UP ON THESE NEW RULINGS?

While Congress has changed the tax laws, the courts have also tinkered. Before you mail off your Form 1040, see how these rulings may affect you. By Sheldon H. Gorlick, J.D. SENIOR EDITOR

You've heard endlessly about all those changes Congress made in the tax laws. Yet other changes with an even bigger potential impact on your income taxes have hardly attracted any publicity. These changes have come from the court decisions that define tax laws. We culled the lawbooks to find the cases most likely to have an impact on the Form 1040 you must file by April 15. Here are the key rulings, each tied into the tax-return section it relates to.

You'll find solutions to such problems as shifting income out of your high tax bracket; making interest deductions stand up for a loan between family members; writing off a casualty loss even though available insurance wasn't claimed; getting shaky car deductions to ride better; strengthening write-offs for entertaining at home; and taking deductions for a home-office that *isn't* for the practice of medicine.

Walter Wick

This story asks: Has your accountant kept himself informed of the latest changes in tax law on allowable deductions? Shown is the tax advisor who is definitely, charmingly not *up with the times. As he works on the 1040 Form, he uses a quill pen and wears a nineteenth-century frock coat. Mr. Pickwick meets Uncle Sam.*

John Newcomb

These illustrations were part of a series used in a general corporate management magazine. The company executive was being urged to be patient with the freelance inventor. Even though they spoke different languages, the executive and the tinkering genius could collaborate to produce a valuable product that bore the stamp of both.

Both of these covers deal with personal problems. The one at right talks about burnout; a candle-wax model of the tired technician was shot using backlighting. The take included frames in which the wick was burning, but this image with the spiraling smoke seemed more eloquent. The cover below discusses sexual harassment. Bringing the camera close called full attention to the woman's scared eyes and the man's wedding ring.

Stephen E. Munz

Developing
an effective
cardiac
profile

A helpful
good-bye:
The exit
interview

Confessions
of a JCAH
inspector

Comparing the
new automated
microbiology
systems

**KEEPING
SEXUAL
HARASSMENT
OUT OF
YOUR LAB**

MLO

MEDICAL LABORATORY OBSERVER • NOVEMBER 1981

**Beating
burnout
in the
laboratory**

A statistical system
of quality control
in hematology

Peer review:
A cost-effective
management tool

Why every tech
should be
a specialist

The employment
interview:
What you can
and cannot ask

Stephen E. Munz (art by Janis Conklin)

Is saving money sometimes a bad idea? This article suggests that a working professional may follow the conventional wisdom and retire with not enough cash in later years, not enough water to fill this dollar-shaped swimming pool.

FINANCES

IS SAVING FOR A RAINY DAY ALL WET?

Inflation, taxes, and disappointing investments have caused some of the best-laid plans to come up dry. Our panelists talk about what they've done right—and wrong—with their money over the years.

MEDICAL ECONOMICS: Retirement planning is high on most doctors' lists of concerns these days. Have you been able to build as adequate a retirement fund as you expected to when you started out?

BATES: It's been a struggle for me to put in enough money to meet the Keogh limits each year. As a solo internist, I haven't enjoyed the large fees that some surgeons have had. Even so, I'd be in an excellent position to retire were it not for inflation. I've invested in the stock market and in money-market funds. But between the income tax bite and inflation, I think I have less money—in terms of purchasing power—than if I had just put the money in a tin box on the shelf. I intend to retire at 65, and I think I'll have enough money. But I haven't anywhere near the affluence some of my colleagues have attained.

EGGER: I'm also solo unincorporated, and I've set Keogh funds aside for myself and my employees. I agree that I probably would have been better off simply putting the money in a tin box. We set aside money for my oldest daughter's education but find that these dollars are nearly worthless today. It would have been wiser to spend

that money on something enjoyable and pay for her college out of current income. That's just what we plan to do with our two younger sons. We're going to enjoy our motor home, take vacations with our kids—use our money as it comes in. After the children are out of the house we'll begin laying money aside for retirement.

MEDICAL ECONOMICS: Other than the Keogh plan, what other provisions for retirement do you have in mind?

EGGER: My best prospect right now is the plan at the hospital where I direct a residency program. I plan to stay at that job unless I'm fired, or the plan should be worth a fair amount. Also, the military is hounding me to get back in again, and what they're offering sounds better and better every day. Maybe even a reserve position might be in order.

MEDICAL ECONOMICS: Have you considered how long you plan to work?

EGGER: Not really. I don't consider what I do to be work. I may never retire. Just one week's vacation drives me insane. There's nobody to stroke my ego, to tell me I'm great.

MEDICAL ECONOMICS: Has anyone else had his retirement planning

thrown off course by inflation?

SHAW: Yes. I found myself in the same position, trying to plan to send my children through college. I started out putting aside a few dollars and thinking that having something paying $1,000 a year would do it. Of course, that kind of money has turned out to be nothing at all. So we've adopted the same philosophy as Dr. Egger—spend out of current income for the things we want and hope that in the time between our children's graduations and our retirement we'll be able to save something. If you try to base a whole spending plan around putting money aside for retirement, you'll find it's almost impossible to do.

MEDICAL ECONOMICS: Dr. Alper, you're the only incorporated physician here. Is your experience largely the same?

ALPER: Unfortunately, retirement planning has disintegrated. In the last few years, all we've seen is unpredictable inflation and gyrations in the value of equities. The fact is, there's no secure investment. You can't plan a retirement program unless you can look into the future and predict what things are going to be worth. As a result, most doctors I know are regarding their ability to do some work after

80 MEDICAL ECONOMICS/FEB 1, 1982

Lonni Sue Johnson

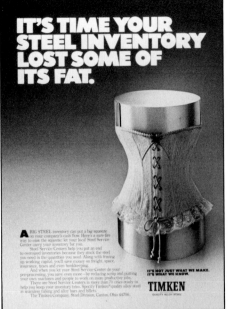

Selling metal and metal products (steel bearings, in the case of the Timken Company of Ohio) can be dull. As the creative supervisor at Batten Barton Durstine & Osborn (BBD&O) said, "Bearings are photogenic but plain. And steel— rolled steel looks like a roll of paper towels." What they did to make Timken steel products memorable was produce a series of elegant analogic images for the ads. These pages got top readership in industry journals and succeeded in portraying Timken as a highly quality conscious company. The headline for the steel rose image at left reads "Anybody can come up with ordinary steels. We come up with hybrids."

These posters for the Milwaukee Museum of Art are designed to appeal to younger people who may think all museums are dull warehouses of dead emotion. The models are good, the puns are engaging, and the copy is lively.

"Art appreciation can make you a bigger person. Join the Milwaukee Art Museum."

Vincent

"Earmark some funds for an art museum membership."

The image of a film reel becoming a snake was crafted for a brochure giving information about an association of professional arts people in Phoenix, The Communicating Arts Group of Arizona. The image alone suggests creativity, drama, and desert—all the ingredients needed to describe this group of young designers.

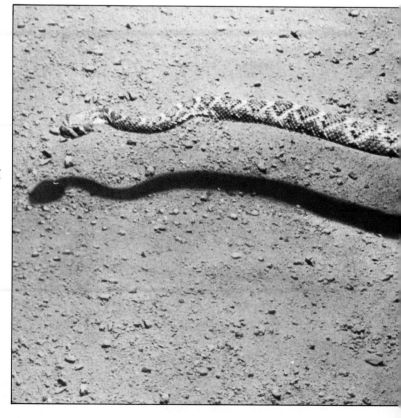

THE BITE SYSTEM: A RELIABLE STARTING POINT

Since our premise is that visual problem solving must start with words, the designer's point of departure has to be the editor's working title or the copywriter's main statement of benefit. The designer uses these words to produce alternate ways to express the same or a related thought, and progresses to parallel ideas that then become counterpoint graphic statements that can be harnessed to the editor's words. Together, the editor's words and the designer's images give the biggest emotional bang for the buck, the maximum wallop that will help lodge the message in the reader's mind.

It is this progression, idea-to-words-to-thoughts-to-words-to-idea, that forms the track of a good designer's creativity. The best point of attack is the first word stage of this cycle: you begin by playing games with the words of the story statement—you pound, push, roll, and knead the words just as a baker works dough.

This is where the Bite System comes into play. It is a word-oriented approach to design, a simple way to get closer to the meaning and spirit of a message and a way to put more graphic "handles" on any statement.

It works this way. You begin your analysis of the problem by dissecting the main terms of the title the editor has furnished or any alternate title that you both feel represents the message. You break the statement down into smaller, more digestible "bites" (this is how the method got its name). Basically, the Bite System is

The start of this race between a tortoise and a hare was shot for a cover that dealt with the problems of picking winners in the stock market. Note that the front feet of both animals are off the ground.

employed to ask the who, what, where, why, and how of the main parts of the title; this helps the designer sort out the different shades of meaning and implications of a message. It is a way to more thorough study —and understanding—of the facts of a problem.

However, the questions used in the Bite System are somewhat more specific than what and how. There are six trigger questions to be asked about the title.

Nature. What is the nature of the subject? This requires a complete and minutely detailed description of each of the main terms in the title. Try to imagine you're explaining this object to a being from another planet. To describe a sausage pizza to someone from another galaxy you would have to list such things as shape, texture, color, ingredients, and taste—all attributes of a pizza that we take for granted would be strange and exotic to such a being. With this standard in mind, then, you write down a full report on the physical appearance of each term; try to list every tiny detail of how it looks and feels.

The remaining five trigger questions may be asked about one central *term* you have isolated by asking the first question, but normally they will be addressed to the central *thought* of the title, the action that's described by the entire statement.

Source. Where does this thing come from? Who or what delivers it? Does it result from a choice made by the person involved? Who or what causes it—an individual, an agency, a company? Is it born or does it just happen as a result of the accidental meeting of two or more elements?

How delivered. Is it sent by mail or given personally? Does it take the form of a command or is it simply a request? Does it result from growth? And from what direction does it come: does it rain from the sky or does it well up through the soil? Does it barge into view with

Another stage of the same race shows two gumdrops scuttling into the sunset. Often a single pose will be planned and executed during a photo session; then the photographer, an assistant, or one of the art directors present at the shooting will suggest a variation that may yield a valuable followup illustration, or perhaps an even better main picture. It is wise to stay flexible and alert to such on-the-spot opportunities.

Walter Wick

great suddenness and violence, or does it seep in unnoticed until there is enough to be seen or otherwise sensed?

Size. What's the space occupied by the subject? This question refers to both its *physical* and *emotional* size; that is, does it loom large compared to a human being or is it microscopic? Is it so great a concern that it crowds everything else from the mind, or is it insignificant and easily forgotten?

Weight. What is its density? This question is also meant to analyze the physical and emotional aspects of the subject's bulk. Is it a crushing burden to carry or is it feather-light and no hindrance at all? Is the thought of it oppressive or can it be easily shed?

Why given. What's the basis in human need, the motivation for giving this thing? What's expected in return? This is a particularly useful question since most editorial or advertising statements are transactional—someone doing something to someone else. Why is it being done? Who stands to gain from the exchange? Do both the giver and the recipient benefit, or is the advantage lopsided? Is the item being sold, or is it bestowed as a favor?

These six questions are asked about the statement, whether it is an editor's working title, an advertising campaign slogan, or an event that becomes the reason for a promotion piece. The designer begins with a message of words, a blank pad, and a pencil. It is very important to write everything down as the Bite System questions are answered; often a word or a phrase will suggest another graphic path to explore. I no longer write these answers and I doubt if any top graphic professionals who have developed their skills intuitively do either—but once in a while a problem won't budge. When this happens I get out my pad of yellow scratch paper and start making lists. In the beginning, though, you should write out everything.

The great virtue of this system is that it is almost totally mechanical. It requires no special equipment; the designer simply writes the title and immediately starts listing attributes of the subject in answer to these six standard questions. A short way into the making of these lists you will become intellectually engaged; your imagination will mesh with the problem factors, and you will be caught up in the simple play activity that exposes one useful solution after another. And it all happens before you have a chance to freeze while looking at a blank pad. There's no time for panic because the first moves are prescribed and very few choices are required. Analysis is begun with a minimum of stress, and literally before you know it you will be well along the way to practical conceptualizing—off to a clean start for getting visual ideas. This system works.

Now for the second stage of the system. You have used the Bite System questions to reach a fuller understanding of the title and its main terms. Next you will paraphrase the main facts from the description. You will use your own words to make a series of statements about the subject; we can call these **source statements.** Each source statement will grow from the more intimate understanding of the original title you will have developed as a result of looking closely at its separate parts by using the Bite System. Each such statement will express a main fact or attribute of the title, a quality you feel is important to consider about the original idea. There may be anywhere from three to six of these alternate or related statements of fact that amplify the original message.

Now the fun begins. We will proceed to play language games with each of these source statements; we'll use word games, phrase games, and fact games. The sentences that result from these language games we will term **substatements,** and it is from these that the visual ideas will evolve.

Word games. These consist of taking the key words in the source statements and substituting synonyms—other words that mean the same thing. We also try inserting antonyms (opposite-meaning words), puns, and double-entendres. Often a substituted term can furnish a shade of meaning that wasn't in the source statement, one that is still true to the spirit of the original title. A good thesaurus is helpful here; I use *Roget's Thesaurus*, one of the first and one of the best.

Phrase games. These make use of known sayings about the subject: proverbs, aphorisms, and bits of traditional wisdom. *Bartlett's Familiar Quotations* is a valuable companion for this part of the probe, a treasure house of the best sayings in human history.

Fact games. These are played by changing one element of the source statement at a time and seeing what happens to the message. This "twist" usually changes the direction of the original title and puts emphasis on the opposite meaning; occasionally the polar opposite meaning in an image that results from a fact game may serve, usually through irony, to enhance the headline statement. Or perhaps standing the source statement on its head, changing the order of the words, will yield an unexpected angle that can be pursued.

Let's see how the Bite System can be applied to some typical assignments. For demonstration purposes, three different types have been chosen: a feature article for a trade journal for doctors, a product development campaign, and a short feature for a consumer magazine.

CASE ONE:

This is an article for medical students graduating in the coming year. It will present, in chart form, information obtained in a nationwide survey that reveals shortages of physicians in some parts of the country and crowding in others. The editors advise doctors who are just starting practice or those who are thinking of relocating existing practices to choose smaller towns in states with less population—not big cities, where competition is tougher. The article says that opening an office in a small town will be easier because the ratio of physicians to patients shows the need for doctors in such places. Doctors tend to flock around the big urban centers—particularly New York and other Eastern cities—partly because the available hospital equipment and technology are more sophisticated. The working title for this story is:

Where your specialty is needed most—and least. The author's chief advice is for the doctor to relocate in a small town, not a large city. Let's apply the Bite System questions to the title.

Nature

where. A small town in any part of the country. A couple of rooms in a small office building with suitable signs that announce a new medical practice, a base for the newly arrived doctor to use, a place where he or she can examine patients.

specialty. A preferred type of medical treatment, some type of medicine for which the doctor has had extensive training in medical school or during his internship. A doctor's job is to correct malfunctions in the human body, to analyze the performance of its parts and prescribe curative drugs, surgery, or other treatment. A doctor may choose any one of dozens of specialties including obstetrics, pediatrics, cardiology, urology, dermatology, and radiology. He may, in other words, choose to treat only certain parts of the human body, and he will strive to develop this expertise. Presumably, a medical student chooses a specialty because it is particularly challenging or because it's needed where he or she plans to practice.

need. People in a community occasionally need treatment and go to their family physician; if there were a need to consult a specialist, the family doctor might refer patients to one in the same town. But if the specialists in a small town are already too busy with their caseloads or if the town has a gap—no one in the given specialty—those patients who need the specialist are in distress; they may be forced to travel long distances to get the help they need. Obviously, both patients and their family doctors would welcome the needed specialists to their town.

Whereas the first Bite System question (nature) is used to dissect the literal terms of the working title, the remaining questions are applied to the main thought of the headline: the need for more specializing physicians in small towns. Again, this change of target, from the terms of the title to the overall message of the headline, is normal in titles that describe or suggest transactions—someone doing something. In such cases it is not useful to apply the last five Bite System questions (source, how delivered, size, weight, and why given) to the separate terms of the headline.

Source. The supply of doctors comes largely from the country's accredited medical colleges and their teaching hospitals. In addition, there are a few established physicians who wish to relocate for personal or family reasons.

How delivered. The town gains a new specialist physician when the doctor decides to settle in the community. An office is opened, a name shingle is hung outside the door, and prospective patients start coming, mostly by referral from general or family physicians who are established in the town.

Size. Where to establish a practice is a question that looms quite large in the mind of a doctor contemplating such a choice. In fact, few decisions will affect his life more drastically.

Weight. And the question will be seen to weigh heavily in the doctor's consciousness. Until the doctor makes a firm decision, thus removing the burden, he or she may be preoccupied with the matter.

Why given. The newly settled doctor can look forward to the esteem and gratitude of new patients and the respect of new colleagues; this is a chance to make a fresh start in life. Secure financial footing will come faster than it would in a big city, and the overall quality of life should improve. Progress from starving student to pillar of the community will take place in a fairly short time, and the doctor can put his or her special training to better use in a place where competition is less severe.

Now let's distill this information into a series of

CASE ONE: Doctors and small towns

1·b *You should set up shop in a small community.*

1·g *A small town is a good place to put down roots.*

source statements, main facts culled from a close examination of the title using the Bite System. We can say:

1. You should be locating your office in a small town.

2. Patients in small towns need more doctors.

3. You'll be happier if you go where you're needed.

4. You may have to travel far to find the best location.

5. You can put your training to better use in small towns.

Then we move to the third stage of the process, playing language games with each of these source statements. First, the word games: watch what happens to the shades of meaning in the first source statement when different words are used.

1. You should be locating your office in a small town.

   a. You should center your practice in a small town.

   b. You should set up shop in a small community.

   c. You should be settling in a village.

   d. You should aim for a small town.

Next, we try phrase games. See what images can grow from using such familiar expressions as putting down roots, staking a claim, establishing a beachhead, pitching camp, and setting up headquarters. These familiar colloquial expressions can yield more statement "handles":

   e. Stake your claim in a small town.

   f. Make your headquarters in a small town.

   g. A small town is a good place to put down roots.

Finally, we play fact games. We can reverse the meaning of the statement, although it seems to have no value in this case.

2-a *There aren't enough doctors in small towns.*

2-c *Go out where a (doctor) friend is a friend.*

h. You should be locating in a big city.

Now let's go to the second source statement:

2. Patients in small towns need more doctors.

Word games:

a. There aren't enough doctors in small towns.

b. There's a shortage of doctors in rural communities.

Phrase games:

c. Go out where a (doctor) friend is a friend. (This comes from a line in a western ballad, "I'm Back in the Saddle Again," which was Gene Autry's radio theme song.)

d. A (doctor) friend in need is a friend indeed.

Fact games:

e. Patients in big cities have more doctors than they need.

f. There are too many doctors in urban centers.

While these statements do not reverse the truth of the original title, they do emphasize the negative factor: the big city.

Let's go to the third source statement:

3. You'll be happier if you go where you're needed.

Word games:

a. Life will be better if you go where you're needed.

b. It's satisfying to find a need and fill it.

c. It's nice to be needed.

d. It's good to go where you'll be appreciated.

2-f   *There are too many doctors in urban centers.*

Phrase games:

e. Find a home in a community that needs you.

f. Make your mark in a small town.

Fact games:

g. You'll be sorry if no one needs your specialty.

Now for the fourth source statement:

4. You may have to travel far to find the right location.

Word games:

a. You may have to cross the country to find the best place.

b. You may have to wander far before finding a good place.

c. Finding a good location may require a lot of shopping.

d. If you persist, you'll find a good place.

Phrase games:

e. Go west, young man (doctor).

f. Seek and ye shall find (a good location).

3-d   *It's good to go where you'll be appreciated.*

4-a   *You may have to cross the country to find the best place.*

Fact games:

    g. A good location may be right on your doorstep (near).

And the fifth source statement:

5. You can put your training to better use in smaller towns.

Word games:

    a. You'll find it easier to compete in a small town.

    b. There aren't as many (obstetricians, for example) in a small town.

Phrase games:

    c. You can be a big frog in a small pond.

Fact games:

    d. Your specialty is too crowded in big cities.

    e. Your training is less likely to be used in a big city.

As you can see, this word munching has produced five **source statements** or major aspects of the original message. Language games used on the source statements have produced no less than thirty-three **substatements,** the specific wordings of which will suggest many graphic metaphors. Here are some of the visual ideas:

- The new doctor's friendly acceptance in a community is symbolized by a small boy selling him lemonade. The boy has just set up shop also. **(1-b)**
- The town is depicted as a supportive landscape in which the doctor can happily put down professional roots, can literally become a pillar of the community. **(1-g)**
- The doctor is shown listening for a heartbeat on a saguaro cactus. **(2-a)**
- The newly arrived doctor is being greeted by a friendly scarecrow. **(2-c)**
- A thicket of handheld surgical scalpels is shown; all of the scalpels are trying to perform the same operation on the same patient at the same time. This refers to the high level of competition in big urban centers. **(2-f)**
- A pleased doctor is shown reacting to many friendly patients' hands reaching for him from offstage. **(3-d)**

4-e   *Go west, young man (doctor).*

4-f   *Seek and ye shall find (a good location).*

• The tyro physician is hitchhiking along a Western highway with his medical bag tied to a pole that rests, hobo-style, on his shoulder. He is wearing a three-piece suit. **(4-a)**

• A primitive-style illustration shows the road out of a city leading to a rainbow arch on the horizon. **(4-e)** Go west . . .

• The new doctor is characterized as Diogenes searching with a lantern. Behind him is the neon-lit city. **(4-f)** Seek and ye shall find.

• The doctor is shown as a big frog locating in a small pond as he attaches his shingle to a lily pad. **(5-c)**

• A stampede of doctors rushes to counsel a pregnant woman patient—another comment on big city professional crowding. **(5-d)**

Look at the idea sketches for a suggested degree of finish for the drawings you make. At this stage only very crude little animated scrawls are needed, just enough description to show the idea. Later, a few of these may be rough-sketched to full size, perhaps with magic marker color or watercolor added.

What have we left when the problem-solving dust settles? Eleven visual ideas that, each in a slightly different way, represent the initial story message: where your specialty is needed most—and least. Some of them will be rejected because they show doctors in undignified or unflattering ways, but most of them can be used as nuclei for visual statements, whether they are finally executed as photographs or illustrations.

Your first visual ideas should result from a free-swinging, uninhibited round of association. No doodle or half-thought should be rejected as absurd or unprofitable. There will be time enough to edit the final images, to expose them to the cold light of day, and test them on other people. It is important to kick up lots of possible choices from the beginning of the analytical process; often a solution that's patently not suitable for the final choice can be modified and made useful. Save them all; they are tender visual seedlings that may grow to graphic magnificence!

5-c   *You can be a big frog in a small pond.*

5-d   *Your specialty is too crowded in big cities.*

CASE TWO:

You are the designer on a creative team assigned to generate a new campaign of print ads for a brand of gourmet-quality chocolate chip cookies. The maker of the product, Famous Amos Inc.* claims it's "the best chocolate chip cookie ever made." The cookies are widely distributed nationally, but the manufacturer promises they are always fresh when bought because they are efficiently packed (and because customers don't leave them on the store shelf long enough to get stale). Famous Amos offers to replace any package of cookies that is found by the purchaser to be stale.

Let's take the central claim by the manufacturer as the working title and apply the Bite System (no pun intended).

Famous Amos has the best chocolate chip cookie ever made.

Here we go with the six Bite System questions:

Nature

best. Most flavorful, most pleasing texture (chewy and soft), satisfying at any time of the day, contains more chocolate pieces than other brands, and is so tasty that eating it is habit-forming.

chocolate chip cookie. A small (two to three inches in diameter) baked cake made with flour, sugar, eggs, butter, and salt; several bits of dark chocolate are poured into the batter before baking. Chocolate is a flavoring substance made from cacao beans imported from South America. The finished cookie is roughly round, golden brown, and lumpy-looking because of the chocolate nuggets. The sooner after baking it can be eaten the better since, if the cookie is left exposed to air, it hardens.

ever. According to Famous Amos, the first recognizable chocolate chip cookie ever made was baked in 1929 in a farmhouse kitchen near Lowell, Massachusetts. However, we may take maximum liberty with the term and use its traditional meaning: for all of recorded history.

made. Manufactured. In this case, "made" refers to the whole process of mixing just the right amount of necessary ingredients, baking the dough for exactly the right length of time in ovens set to produce the right amount of heat, and, finally, cooling and packaging. This is a finely tuned, carefully controlled process that must be repeated each time to produce cookies of sufficient quality to carry the brand name.

In this case, the main slogan is not transactional; it does not describe anything the cookies are doing. It just makes a claim for what they are: the best ever. Therefore, the other five Bite System questions may be used to analyze the main term of the slogan: cookies.

Source. The cookies begin their commercial journey in several regional Famous Amos bakeries.

How delivered. Normally by truck. The packages are carefully hauled from the branch bakeries to retail stores and gourmet shops nearby. This is a regularly scheduled delivery.

Size. Physically, the cookie is fairly small, easily held in the hand. Emotionally, the cookie may grow in the consumer's mind as his taste for the cookie grows.

Weight. Again, the physical density of the cookie is slight—a fraction of an ounce—but the urge to eat another may be fairly strong psychologically and may border on compulsion.

Why given. A person may eat the cookie to satisfy his own appetite or at least to blunt his hunger pangs. He may give it to another person for a variety of reasons: to reward performance, to be remembered as a giver of

---

*Use of the brand name is by courtesy of Famous Amos Inc. This demonstration does not obligate the company to support any advertising claims made as part of the exercise.

gifts, to serve as bait to induce action. Or the cookies may simply be dessert at the end of a meal.

From this breakdown of the original statement we can make five source statements or related facts about the slogan:

1. Famous Amos cookies have always been the best.

2. The cookies are always fresh and chewy.

3. You can—and should—take them any-where.

4. Famous Amos cookies are addictive.

5. Famous Amos cookies are works of art.

And again, let's play language games with these. The first source statement is:

1. Famous Amos cookies have always been the best.

Word games:

a. Famous Amos cookies have been the best throughout history.

b. Famous Amos cookies have had a great record in the past.

c. Famous Amos cookies have kept high standards.

Phrase games:

d. Famous Amos has always been number one.

e. Famous Amos—always on top.

f. Famous Amos, the superstar of chocolate chip cookies.

Fact games:

g. You don't want a Famous Amos cookie—if you have no taste.

CASE TWO: Famous Amos cookies

1-a *Famous Amos cookies have been the best throughout history.*

1-a *Famous Amos cookies have been the best throughout history.*

Next, the second source statement:

2. The cookies are always fresh and chewy.

Word games:

  a. The cookies are always ready for you.

  b. The cookies are always soft.

  c. The cookies seem as though they're just out of the oven.

Phrase games:

  d. These cookies are not so tough.

Fact games:

  e. These cookies are rarely fresh.

  f. Tough cookie!

Now for the third source statement:

3. You can—and should—take them any-where.

Word games:

  a. You can take them skydiving, skiing, roller-skating, to the opera, etc.

  b. Never get caught without your cookie.

Phrase games:

  c. Have cookie, will travel.

  d. There's a Famous Amos cookie for every occasion.

Fact games:

  e. Famous Amos cookies are so good they should be kept at home.

  f. You can never keep Famous Amos cookies in the house.

1-d *Famous Amos has always been number one.*

1-b *Famous Amos cookies have had a great record in the past.*

On to the fourth source statement:

4. Famous Amos cookies are addictive.*

Word games:

a. You can't stop eating Famous Amos cookies.

b. One Famous Amos cookie leads to another—and another.

c. You must keep eating Famous Amos cookies.

d. You could get carried away eating Famous Amos cookies.

e. Once you have one you can't live without them.

Phrase games:

f. You may have a (cookie) monkey on your back.

Fact games:

g. Who needs another cookie—even a great one?

h. A cookie is the last thing I need.

Finally, the fifth source statement:

5. Famous Amos cookies are works of art.

Word games:

a. Famous Amos—the masterpiece cookie.

b. The Famous Amos cookie: eighth wonder of the world.

c. Man's greatest works: the wheel, the printing press, and the Famous Amos chocolate chip cookie.

Phrase games:

d. A Famous Amos cookie is one of a kind.

e. Famous Amos, the Picasso of cookies.

2-b *The cookies are always soft.*

1-f *Famous Amos, the superstar of chocolate chip cookies.*

2-f *Tough cookie!*

---

*This is a bad connotation for a cookie—especially bad for a chocolate-flavored cookie. No manufacturer would like to hint that it sells munchies laced with drugs. But let's not let this concern inhibit getting visual leads. We can use the clue of a craving or acquired taste for the cookie and still be respectable.

Fact games:

f. It's just a humble cookie.

g. It's only a chocolate chip cookie—and the Alps are only mountains.

Now we have thirty-four **substatements** distilled from five **source statements** that were, in turn, derived from the advertiser's message by using the Bite System. Let's see what images they suggest:

• A Roman charioteer is munching a bite taken from his chariot wheel; the wheel is a large chocolate chip cookie. **(1-a)** The best throughout history.

• Marie Antoinette flings chocolate chip cookies to the Paris mob. "Let them eat cookies!" **(1-a)**

• The Union Pacific and Central Pacific railroads link lines in Utah. Engineers are reaching forward to fit two halves of a cookie as they meet. **(1-b)** A great record in the past.

• A victorious prizefighter has his arm held high in victory. His head is a cookie. **(1-d)** Always has been number one.

• Supercookie flies high over the skyscrapers of Metropolis. **(2-b)** The superstar of cookies.

• A man sleeps peacefully with his head resting on a large chocolate chip cookie instead of a pillow. **(2-b)**

3-a *You can take them skydiving.*

3-a *You can take them roller skating.*

Always soft.
● A defiant cookie postures as king of the snack mountain. This one is a ba-a-ad cookie! **(2-f)**
● A large chocolate chip cookie and a man are skydiving together. Each has his own parachute. **(3-a)**
● Rollerskates are shown close up, two with human feet and one carrying several animated (happy) cookies

bobsled-style. **(3-a)** You can take them anywhere.
● A nervous man tries to hide his nudity behind a large chocolate chip cookie. **(3-b)** Never get caught without your cookie.
● A knight in armor sports a large cookie for a shield. **(3-c)** Have cookie, will travel.
● A chocolate chip cookie zips across the page in a full run. Its feet have winged Mercury slippers. **(3-f)** You can never keep these cookies in the house.
● A line of chocolate chip cookies walks from left to right; all are holding hands. **(4-b)** One leads to another —and another.

3-b   *Never get caught without your cookie.*

3-c   *Have cookie, will travel.*

3-f   *You can never keep Famous Amos cookies in the house.*

4-b   *One Famous Amos cookie leads to another—and another.*

- An old sourdough drags himself across the desert; he dreams of a chocolate chip cookie. **(4-e)** Once you have one you can't live without them.
- Vincent van Gogh is shown working on his sunflower still life; the centers of the flowers are cookies. **(5-a)** The masterpiece cookie.
- Leonardo da Vinci paints the Mona Lisa, seen munching a cookie. **(5-a)**
- Pablo Picasso works on a cubistic portrait of a woman; her cheek is a cookie. **(5-a)**
- A monumental cookie is balanced on edge at the top of the Great Pyramid of Cheops. **(5-b)** Eighth wonder of the world.
- A museum displays a large Magritte-type super-realistic image of a huge cookie floating in midair above a landscape. A man and a woman are shown admiring the painting. **(5-d)** One of a kind.

The final tally: nineteen visual ideas. A review of these doodles shows possibilities for campaign series with two general themes:

1. The product can be fit into historic vignettes. It can serve as a chariot wheel at the dawn of the mechanical age; it may represent the largesse of Queen Marie Antoinette of France; or it can symbolize the joining of the transcontinental railroad in 1869. There are good slogan opportunities here: "Famous Amos Moments in History," perhaps. Famous Amos cookies can be shown to play a part in the campaigns of Napoleon (now you know why Napoleon is usually shown with his hand hidden in his tunic; this is where he kept his private

5-a   *Famous Amos—the masterpiece cookie (Leonardo).*

4-e   *Once you have one you can't live without them.*

stash of cookies). Or perhaps an Egyptian slave was given a Famous Amos cookie break once a day by an unusually kindly overseer as they worked to finish the Pharaoh's tomb. Or maybe the Famous Amos chocolate chip cookie was a favorite of Queen Victoria as she and Prince Albert discussed the affairs of empire at high tea each afternoon. The string of playful scenes based on great historical events or personages is endless—amiable, tongue-in-cheek nonsense—but an entertaining way to promote the product nonetheless.

2. The cookies can be shown as props in great paintings that are being completed by the masters: van Gogh, Leonardo, Picasso, Goya. This suggests a theme slogan, perhaps a small repeating overline for each ad, such as

"Famous Amos, the Masterpiece Cookie." This slogan allows plenty of elaboration on the subject of quality control in the copy; the theme—meticulous care in preparation—harmonizes with the main assertion that each Famous Amos cookie is a work of art.

Some of the other substatements that were not developed are probably worth more attention. Try making thumbnail sketches for these:

(1-g) You don't want a Famous Amos cookie—if you have no taste. Draw your own version of the Beatles'

5-a   *Famous Amos—the masterpiece cookie (Van Gogh).*

5-a   *Famous Amos—the masterpiece cookie (Picasso).*

Nowhere Man—a totally tasteless, unaware cipher of a fellow whose attire is a tailor's nightmare.)

(3-d) There's a Famous Amos cookie for every occasion. (Weddings, graduations, bar mitzvahs—imagine all the participants at such an occasion eating cookies during the ceremony.)

(3-e) Famous Amos cookies are so good they should be kept at home.

(4-g) Who needs another cookie—even a great one? (The consumer is dieting and resolves to refuse dessert; he is won over by a Famous Amos cookie. The subhead at the bottom of the ad might answer the headline question of "Who needs it?" with "I do.")

(4-h) A cookie is the last thing I need. (If this line were put in the context of a firing squad, the statement would gain new color. Asking for a last cookie instead of a last cigarette is probably healthier anyway.)

(5-g) It's only a chocolate chip cookie—and the Alps are only mountains (and the Grand Canyon is only a riverbed, baseball is only a game, etc.).

The associative possibilities are nearly endless. There are many ways to help the customer find a special relationship to the product—the cookie—and to entertain on the way.

5-b　*The Famous Amos cookie: eighth wonder of the world.*

5-d　*A Famous Amos cookie is one of a kind.*

CASE THREE:

You work for a national publication that deals with tennis and tennis players. Your editor has a story about opponents who consistently call every ball "out" that comes to their side of the net and lands anywhere near a court boundary line. Many times the intentional miscall is obvious but it is futile to argue. This person—and he or she can be a crony or a total stranger in a pickup match—may even yell "Out!" before the ball bounces. This is a highly annoying habit that can ruin your game. The editor's title is:

How to protect yourself against the tennis cheater.

Once again we'll use the Bite System questions on the message.

Nature

protect. To shield yourself from injury or insult. In this case the need is to take precautions or adopt effective countermeasures.

the tennis cheater. As defined above, this person warps reality in his or her favor when playing tennis. The cheater is prompt and adamant about line calls on his or her side of the court and they are always made to the cheater's advantage—your ball is always long and you lose the point. This happens with such frequency and regularity that you can be sure the opponent is knowingly calling balls out that landed in bounds.

In this case, the main topic of the title is cheating at tennis. The other five questions may be applied to either the main term in the title, the tennis cheater, or the dominant topic, cheating.

Source. The originator of the problem is your opponent.

How delivered. The opponent shouts to you across the court, yelling "Out!" or "Long!" or, if a serve, "Fault!"

Size. The only meaningful observation on size is that the oppressiveness of your opponent's habit can fill your mind to the detriment of your game; it can crowd out needed space for play strategy.

Weight. Substantially, the comment is the same for weight as it is for size: this flagrant abuse of the rules may be increasingly heavier the longer you play with this opponent.

Why given. To win . . . and to win at any cost. Obviously such an opponent does not care about your opinion or fear damaging his or her own self-respect. The desire to win nearly reaches the level of blood lust, and anyone who stands in the way of victory will get trampled. This kind of player is compulsive and has personal motives: to advance higher on the club ladder and acquire the reputation of being one of the club's best players.

Let's see what kind of source statements this yields. We can say:

1. You play tennis with a cheater.

2. He always makes close calls in his favor.

3. He makes line calls quickly.

4. This person will do anything to win a game.

5. Your opponent's habit is annoying to you.

6. There are countermeasures you can take.

Let's proceed with language games on the first source statement:

1. You play tennis with a cheater.
Word games:
   a. Your opponent uses shady practices.
   b. You're playing with a crook.
   c. Your foe is dishonest.
   d. It's not possible to play an honest game with this person.
Phrase games:
   e. You are victimized by your opponent.
   f. You get taken every time.
Fact games:
   g. You don't play tennis with this person. (He's playing something else.)
Now for the second source statement:

2. He always makes close calls in his favor.
Word games:
   a. He always says the ball you hit is out.
   b. You can't argue with this opponent.
   c. You can't win against this person.
Phrase games:
   d. Your opponent loads the dice to favor himself.
   e. Playing tennis with this person is a no-win proposition.
Fact games:
   f. He never calls them close.
   g. He wouldn't take advantage of you.
Let's go to the third source statement:

3. He makes line calls quickly.
Word games:
   a. He shouts "Out!" soon after you hit.
   b. He calls a ball out before it bounces.
   c. He says a ball is out even if he hasn't seen it bounce.
   d. How does your ball always land outside the line?
Phrase games:
   e. He makes lightning line calls.
   f. Fastest tennis mouth in the West.

Fact games:
   g. Your opponent is slow to call balls on his side.
The fourth source statement:

4. This person will do anything to win.
Word games:
   a. Your opponent will stop at nothing to win.
   b. Nothing is beneath your opponent.
   c. No tactic is too low for your opponent to use.
   d. Your opponent is completely unscrupulous.
Phrase games:
   e. Your foe would kill to win.
   f. Your opponent would sell his soul to win at tennis.
Fact games:
   g. This person is easygoing when he plays.
   h. This opponent is a pushover.
The fifth statement:

5. Your opponent's habit is annoying to you.
Word games:
   a. Your opponent's close calls distract you.
   b. You resent always hearing that your ball is out.
   c. You hate not being able to trust your opponent.
   d. You wish there could be an impartial line judge.
   e. Anger at his calls spoils the game for you.
Phrase games:
   f. Your opponent is a pain in the neck (or net).
   g. His close calls make you play under a black cloud.
   h. His calls make you see red most of the time.
Fact games:
   i. You could care less about your opponent's calls.
   j. Your opponent's biased line calls really make no difference.
And finally, the sixth source statement:

6. There are countermeasures you can take.
Word games:
   a. You can protect yourself from his tactics.
   b. Take shelter from bad calls by your opponent.
   c. Bad line calls needn't upset you.

Phrase games:

d. There are ways to police your opponent's game.

Fact games:

e. There's nothing you can do to combat his calls.

This gives us a net total of forty-four substatements drawn from six source statements. Let's see what visuals we can get from these.

● A card shark, a dealer with green eyeshade and garters on his shirt-sleeves, is shown at the end of a down-size tennis court that has been made by repainting a ping-pong table. Poker chips and tennis balls rest on the "court" next to the dealer. **(1-a)** Shady practices.

● A masked bandit leans on the end of the ping-pong table "court." He waits with a sly and knowing smile. **(1-b)** Playing with a crook.

● The cheating opponent is shown using a skeet rifle to shoot down your ball before it can bounce in bounds. **(1-g)** Tennis isn't the game he's playing.

● The opponent is shown as a huge grinning face painted on a wall that is preventing any of your shots from landing in his court. **(2-c)** You can't win against this person.

● Your opponent is about to serve; you are supposed to receive serve but it can be seen that your hands are tied. **(2-e)** A no-win proposition.

● The opponent raises a flag that says "Out" as your ball

CASE THREE: The tennis cheater

1-a *Your opponent uses shady practices.*

1-b *You're playing with a crook.*

1-g *You don't play tennis with this person.*
    (He's playing something else.)

bounces, but he has a brown paper bag over his head. **(3-c)**

• The opponent is shown as a wolf in sheep's clothing. He stands, holding his racquet and grinning, ready to miscall your serve. **(4-c)** No tactic is too low to use.

• The foe sees your ball bounce inside the line but holds a placard for you to see; it shows a ball landing past the line. **(4-c)**

• The unscrupulous opponent is shown waiting for serve while standing on the back of a prone little old lady, presumably his own mother. Or grandmother. Or any-one else's grandmother. **(4-d)**

• The opponent is waiting for serve. He's fitted with horns, a tail, and hooves. **(4-f)** He would sell his soul to win.

• Blind justice holds a balance scale and a rule book. You, the reader, are on one half of the scale, your lying foe is seen pointing at the ground on the other half of the scale. **(5-d)** You wish for an impartial line judge.

• Your opponent's biased line calls are getting you very upset. You, the reader, are shown in a mirror, and steam is escaping from the neck of your shirt. **(5-e)**

2-c *You can't win against this person.*

2-e *Playing tennis with this person is a no-win proposition.*

• Your opponent is seen through the lenses of your sunglasses and the image is monochromatic—blood red. **(5-h)** Close calls make you see red.

• You, the receiver, are deep in a foxhole at one end of the tennis court and you wear a steel helmet. Your opponent, a tiny figure at the other end of the court, is lobbing a heavy rain of tennis balls, mortar shells, and bombs at your foxhole behind the base line. **(6-b)** Take shelter.

This gives us a final crop of fourteen visual paraphrases of the original title, "How to protect yourself from the tennis cheater." The solution that was actually chosen was the card shark. There were two main approaches to this topic: we could have shown the iniquity of the cheating opponent, or we could have concentrated on the plight of the reader, the person being victimized. The magazine chose to represent the cheater.

You can see from the discussion of these three disparate design problems—a trade journal feature, a print campaign, and a special-interest consumer magazine article—that the Bite System can be used to analyze

most kinds of conceptual problems. Further evidence of its usefulness can be seen at the end of this chapter; several of the concept covers produced for *Medical Economics* magazine as well as alternate ideas are reproduced to further demonstrate the method.

The Bite System will help art directors, designers, and illustrators most, but copywriters and editors—so-called wordsmiths—would do well to study it, too. These people must deal with graphic designers and should understand their methods. The modern creative team needs editors and writers who are visually articulate and graphics professionals who can use words: if

4-c *No tactic is too low for your opponent to use.*

3-c *He says a ball is out even if he hasn't seen it bounce.*

4-d *Your opponent is completely unscrupulous.*

this mix exists and both types of creative professionals blend their efforts, higher-quality communication is sure to result.

The system may seem cumbersome to you at first. Yes, it will take time to master, but doing anything well takes time. Once you get so familiar with the approach that you internalize the steps—that is, when you can ask yourself the questions and don't have to write the answers—it will be seen that its chief benefit is to condition your mind to tackle a problem this way automatically. As you learn the method and invest this kind of time exploring a given problem, chances are you'll produce a solution—perhaps several solutions—that are effective and surprising, too. If you devote the necessary time and effort, you will probably get the results you want. Of how many pursuits in life can that claim be made? Count them on either hand.

4-c *No tactic is too low for your opponent to use.*

4-f *Your opponent would sell his soul to win at tennis.*

5-h *His calls make you see red most of the time.*

5-e *Anger at his calls spoils the game for you.*

5-d *You wish there could be an impartial line judge.*

6-b *Take shelter from bad calls by your opponent.*

HOW TO PROTECT YOURSELF AGAINST THE TENNIS CHEATER

Don't let the court chiselers take advantage of you. Learn how to spot the seven main types and how to deal with them.

BY BARRY TARSHIS
Contributing Editor

A few years back, a certain tennis player in town invited me to his country club to play tennis, have lunch and then take a swim. I accepted the invitation even though the guy had a reputation for cheating that was second-to-none in the tennis circles I run in. I did it for two reasons. One, at that stage in my tennis career I had yet to play with anyone who had made a legend of himself as a cheater. And two, I had never been to this country club and was curious to see if it was as nice as everybody said it was.

As it turned out, the club *wasn't* as nice as everybody said but my host for the day more than made up for this credibility gap by being a much worse cheater than I had been led to believe. There was nothing subtle about his approach. He started issuing bad calls in the very first game (you would think that since we hadn't played before, he might have at least *waited* a game or two before he started in!), and I don't know for the life of me how I managed to be serving for the set at 5-4.

Which is when it happened—that is, the specific call

PHOTO BY JEFFREY FOX   TENNIS/March 1979   71

Jeffrey Fox

*The solution used was **1-a**, the opponent who uses shady practices. He was portrayed as a card shark with a clearly unsavory look. The "court" was a made-over Ping-Pong table, and the model kept up a patter that would have done justice to a midway drummer—"Just one more time, sucker."*

THE BITE SYSTEM AT WORK

The series of Medical Economics cover ideas shown on the following pages best demonstrates the versatility and idea-producing power of the Bite System. The final design in each example was the result of a search for nuances of meaning in the original statement. Occasionally a striking visual caused a different verb or adjective to be added to the final title, but in each instance the starting point was the same: the editor's message. Alternate ideas were presented in sketch form to the editors and for this demonstration, each of these sketches has been keyed to the Bite System question that produced it. This key appears in parentheses at the end of each description.

PROBLEM

At the end of this calendar year it was discovered that the consumer price index rose four percentage points more than the income of the typical doctor. The inflation rate and the cost of living were punishing everyone, doctors included.

A *The physician is shown hanging precariously from a trapeze. This was rejected because his income, at a national median of $86,000 at the time, wasn't very precarious. (Nature)*

B *The little doctor is blissfully unaware of what inflation is about to do to him. (Size)*

C *Poor economic protection from the eroding rain of inflation is afforded by the ruined umbrella. This symbol is accurate, but the image is probably more useful for a story about pension plans. (How delivered)*

D *The doctor gets dumped by the economic seesaw. (How delivered)*

SOLUTION

The chosen image shows the doctor holding his shin and grimacing with pain. His physician's bag is flying open and pills are exploding from it. The final title, "How's inflation treating you?" leaves little doubt about the answer furnished by the picture. So fertile was this subject area that, after the main cover shot was produced, several variations were staged. Other situations that were used on inside pages showed our model physician reacting indignantly to a wind-robbing blow to the belly, a rap on the head, a punch on the upper arm, and a boot in the backside. The model claimed that he hurt for days from the imaginary bruises on his body. (How delivered)

"We're borrowing to meet our payroll!"

How medical families spend their money

Did these doctors blacklist malpractice patients?

SPECIAL REPORT

Using the new tax law

SEPTEMBER 28, 1981

medical economics

HOW'S INFLATION TREATING YOU?

A Continuing Survey report on doctors' earnings

Walter Wick

PROBLEM

The cover story described the increasingly difficult plight of small medical clinics and offices such as those run by the magazine's typical readers. The competitive pressure was being applied by nationwide chains of hospitals that were being run aggressively as businesses. The working title was "How for-profit hospitals are going after your patients."

A *The large, voracious chain hospital is shown as a giant vacuum cleaner that sucks patients from the whole community. (Nature)*

B *The hospital is shown as a bullying intruder disturbing the tranquility of the town. (Size)*

C *Big fish swallow little fish, so why couldn't big hospitals gobble up little clinics? (Size)*

A

B

C

SOLUTION

Hungry hospitals furnished the final idea. Video games were the pop culture rage at the time this cover was done, so the magazine editors decided to echo the action of these games by symbolizing large hospitals swallowing rows of symbolic patients. The style of the drawing and the addition of a little motion blur made the allusion so clear that no one would miss the point—unless he or she had been living in a cave the last few years. (Nature)

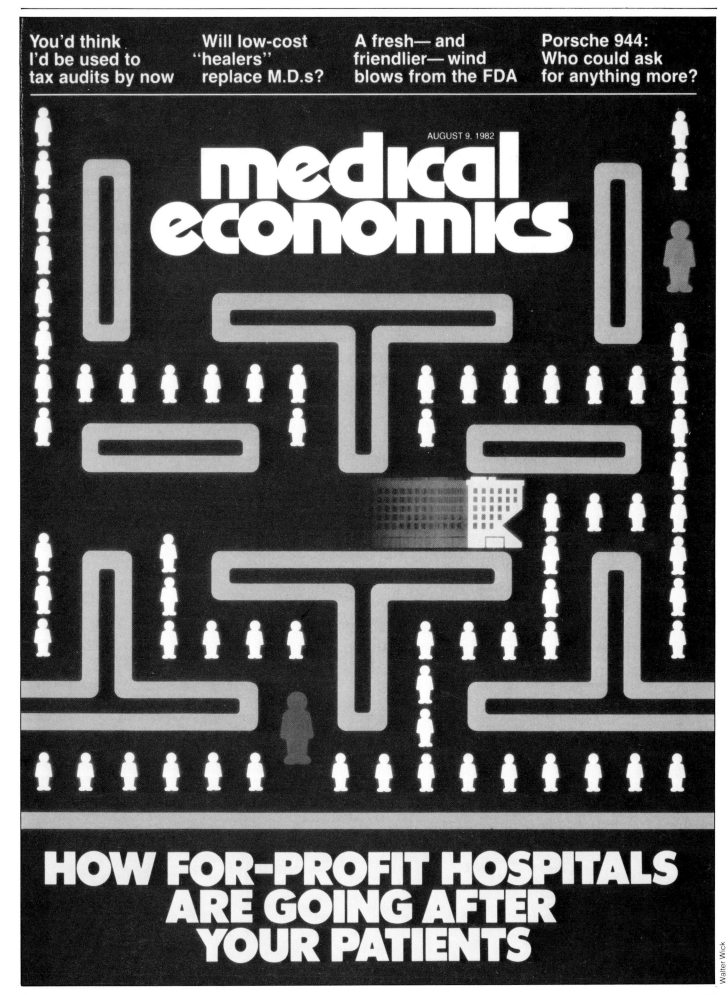

You'd think
I'd be used to
tax audits by now

Will low-cost
"healers"
replace M.D.s?

A fresh— and
friendlier— wind
blows from the FDA

Porsche 944:
Who could ask
for anything more?

AUGUST 9, 1982

medical economics

HOW FOR-PROFIT HOSPITALS ARE GOING AFTER YOUR PATIENTS

Walter Wick

PROBLEM

*Election years traditionally
have brought good opportunities
for alert investors. This issue,
published at the beginning of
1984, featured a lead story
entitled "Picking investment
winners in an election year."
Exploration focused on two
main treatments: first, the Wall
Street symbols of bull and bear
could be shown in a political
situation or, alternatively, the
political party symbols,
elephant and mule, could be
shown in a* financial *scene.*

A *The Republican elephant
and the Democratic mule watch
an electronic ticker-tape report.
(Source)*

B *The elephant and the mule
are shown engrossed in the
financial pages of a newspaper.
(Source)*

C *The legs of the Wall Street
bull and bear are shown in
adjacent voting booths. (Source)*

A

B

C

SOLUTION

*The idea chosen shows a bull, the symbol of a stock
market in which most investors are making
money, as a successful political candidate. The
spotlight is on him as he raises his hooves in
triumph and confetti and ticker tapes fill the air.
The image is upbeat and the picture is
wonderfully executed. (Nature)*

Your
1984 tax
calendar

Hospital politics
won't let me stop
a dangerous doctor

How I beat
Blue Shield's
gestapo tactics

SPECIAL
REPORT
Washington's 1984
game plan for physicians

JANUARY 9, 1984

medical
economics

PICKING
INVESTMENT WINNERS
IN AN ELECTION YEAR

Walter Wick (art by Kathy Jeffers)

PROBLEM

As it does most years, Congress made adjustments in tax laws for 1983 that were particularly hard on the high-earning professional. Our magazine editors felt that readers needed special advice to minimize the shock of the new legislation. The article was entitled "What the tough new tax laws mean to you."

A This solution emphasized cooperation between the government and the taxpayer, a sharing of plans. The doctor and Uncle Sam would be riding a tandem bicycle. (Nature)

B Uncle Sam could be shown as the doctor's accountant, another way of symbolizing cooperation. (Nature)

These first two solutions were judged to be off target since the article suggested more of an adversarial relationship.

C Now Uncle Sam is riding the doctor/taxpayer. The new tax rules are seen as a burden. (Source)

D The doctor is entering a new maze of financial decisions that prevents him from attaining maximum tax benefits. (Nature)

A

B

C

D

SOLUTION

The maze solution suggested other games, other ways to match wits with the IRS. The chess analogy was the next logical step and the editors agreed to alter the headline to support the visual; the title became "Financial strategies that counter the tough new tax rules." Chess pieces were carved to represent the physician and Uncle Sam, and the stage was set for fiscal battle. (Nature)

His malpractice coverage fell $2.4 million short

Surprise! Maybe you shouldn't borrow on your life insurance

You don't need gimmicks to boost your practice

Investment advisers who won't go down with the Dow

JULY 25, 1983

medical economics

FINANCIAL STRATEGIES THAT COUNTER THE TOUGH NEW TAX RULES

Stephen E. Munz (art by Al Pisano)

PROBLEM

The assignment was to dramatize information from a national survey about the resale value and depreciation rate of new automobiles. The working title was "What car gives the best value for your money?"

A A chrome hood ornament in the shape of a piggy bank was one way to suggest worth. (Nature)

B A car made of money, one with a solid gold grill and dollar motifs in the design, spelled out value—but it could also have meant high cost or expensive maintenance. (Nature)

C An extremely cautious prospective buyer is shown examining the car with a jeweler's glass. (Why given)

D Doctor marries car: a wedding, the symbol of long-term commitment and high hopes for mutual support, was suggested. (Why given)

A

B

C

D

SOLUTION

The editors wanted to return to the aspect of thrift. It was simply smarter to buy a car that would conserve its worth longer, so the graphic solution used this specially built piggy bank on wheels. The tiny headlights and taillights were wired and lit for the final photograph. The title was adjusted to harmonize with the picture. (Why given)

What
dying patients
fear most

I faced 15 years
in prison on
trumped-up charges

"I haven't
paid any taxes
in 12 years"

Is any attorney
worth a
$4.4 million fee?

FEBRUARY 7, 1983

medical economics

WHICH CAR
WILL GIVE
YOUR MONEY
THE BEST
RIDE?

Stephen E. Munz (art by Kathy Jeffers)

PROBLEM

The New Jersey state medical board had processed almost twice as many complaints of malfeasance as had any other state during the year past. These complaints, issued against individual practitioners, were turned over to the state civil authorities for investigation and possible prosecution. The magazine editors felt that the state medical board was, for whatever reason, being over-zealous. The title was "Has doctor policing gone too far?"

A *The individual doctor is seen dodging a rain of lightning bolts hurled by the medical board and its cohorts in the courts. (Nature)*

B *The lone doctor is being subjected to the third degree at the police station. (Source)*

C *The doctor is tied to the path of an oncoming (judicial) train. (Nature)*

D *A huge tank trains its main cannon on the individual doctor. (Nature)*

SOLUTION

The notion of overkill, the idea of using too much firepower to cope with the problem at hand, emerged. The final design showed blind justice with a bazooka; a very patient secretary from the magazine's office held still for three hours in order to be covered with white greasepaint. (Nature)

Melvin Belli:
what stops me from
suing a doctor

Why your retirement
plan may take
a great fall

New
inroads into
confidentiality

Which discount
broker gives you
most for the least?

AUGUST 23, 1982

medical economics

HAS DOCTOR POLICING GONE TOO FAR?

PROBLEM

This was a story about a doctor and his associates caught in a true no-win situation: whether they agreed or refused to treat the patient, the case would have led to litigation. The working title was "Whatever we did for the patient we knew we'd be sued."

A *The first solution grew from a fine old expression; the doctor is caught between a rock and a hard place. (Nature)*

B *The doctor is shown on a pancake griddle over a fire. (Nature)*

C *The doctor is tied to a stake, ready to be roasted. (Nature)*

D *Accusations come from all sides. (Source)*

E *The doctor is in a high-risk predicament; he's the assistant in a carnival knife-throwing act. (Nature)*

A

B

C

D

E

SOLUTION

Walking the plank seemed the ultimate form of pressure anyone could experience. It was also a viable symbol for lack of choice; the doctors could jump and be eaten or walk back to face the knives. The final photograph simplified the sketch concept in that one doctor represented the group, the sharks were implied rather than shown, and a large scalpel served in place of a cutlass for inducement. (Nature)

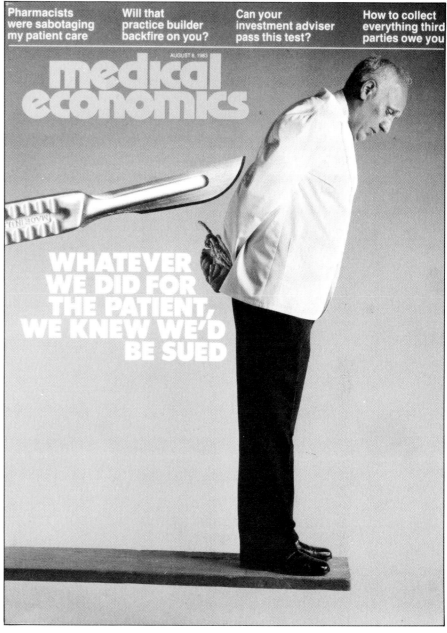

Stephen E. Munz

PROBLEM

The lead story asks, "Are you paying more taxes than you should?" This irresistible question deserved a vivid image and the search was directed toward showing ways in which money could be wasted.

A *The doctor's hand is shown launching a paper airplane made from a folded piece of paper money. An enlarged version of a bill could have been made to allow the character of the missile to be easily seen. Literally, this would show money being thrown away. (Nature)*

B *A blindfolded physician is shown filling out his tax report. (Nature)*

C *The doctor pours his dollars down a funnel. (Nature)*

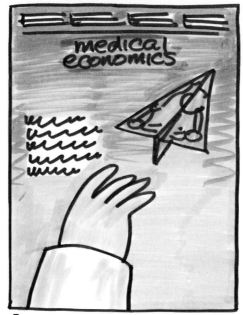

A

B

C

SOLUTION

The notion of continuous waste led to the idea of burning money, of shoveling it into a fire or furnace. A convincing federal furnace (or at least a plastic façade) was made and a backlit transparency of a fire was taped to fit the door opening. An added touch of realism was achieved when the photographer brushed black dry tempera powder around the hole to simulate soot stains. In the final image, our unhappy doctor is shoveling loads of real cash into the flames (and the image was made without violating any city fire ordinances). (Nature)

DECEMBER 6, 1982

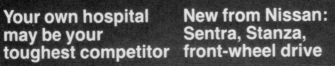

medical economics

ARE YOU PAYING MORE TAXES THAN YOU SHOULD?

Walter Wick (art by Manhattan Model Works)

PROBLEM

Another chapter in the continuous war between the Internal Revenue Service and the individual taxpayer. If, as most of the magazine's readers do, you itemize deductions, there is an ever-present risk of being audited. It's not known with any degree of certainty what brings the federal eagles down on a given report, but recent claim patterns that have attracted audits give some clues about which deductions will or will not be accepted. The working title was "How you can become a smaller target for a tax audit."

A *The doctor has obviously claimed the wrong deduction; he's caught in a snare. (How delivered)*
B *The doctor is protected from IRS arrows by a helmet. (How delivered)*
C *The physician takes shelter beneath an outsized 1040 Form. (Why given)*
D *Doctors become cut-out tin targets in the IRS shooting gallery. (Nature)*

A

B

C

D

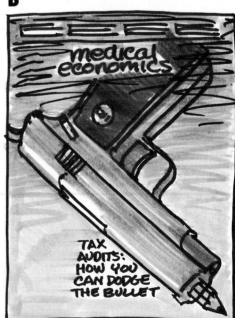

SOLUTION

The picture of an IRS gun and the expression "dodging the bullet" came together to produce the final statement. A pistol replica was modified by a prop artist to allow it to fire a different kind of lead. (How delivered)

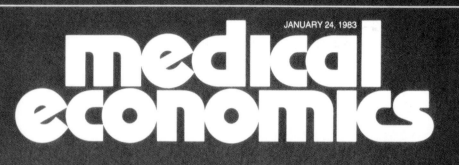

Answers to your tax questions

" I'm retiring on $50,000 a month. Where did I go wrong?"

Forms that make insurers pay up

Melvin Belli: how doctors get sucked into malpractice suits

JANUARY 24, 1983

medical economics

P.38

m.g.c. 67

IRS

TAX AUDITS: HOW TO DODGE THE IRS BULLET

PROBLEM

Free-standing medical clinics that stay open day and night are the new competitive threat to the office-based physician. These curb-service treatment centers—the fast-food chains of health care—arouse considerable anxiety among the readers of the magazine.

A *A surgeon on skates is shown bringing health care to the patient in the same way a carhop would deliver food to customers at a drive-in hamburger stand. (How delivered)*

B *A McDonald's-style neon sign over the clinic door boasts of high customer volume and implies the low cost of treatment that induced so many patients to use their service. (How delivered)*

A

B

SOLUTION

The concept of fast processing gave birth to using the assembly-line, car wash analogy for the cover picture. The artist's model shows a continuous line of patients being run through the clinic. Upon emerging, the people immediately jump up and walk away. The image says that this kind of treatment is short on individual involvement between patient and doctor, but that it has appeal for a growing number of people. It is a threat to the reader's practice. (Nature)

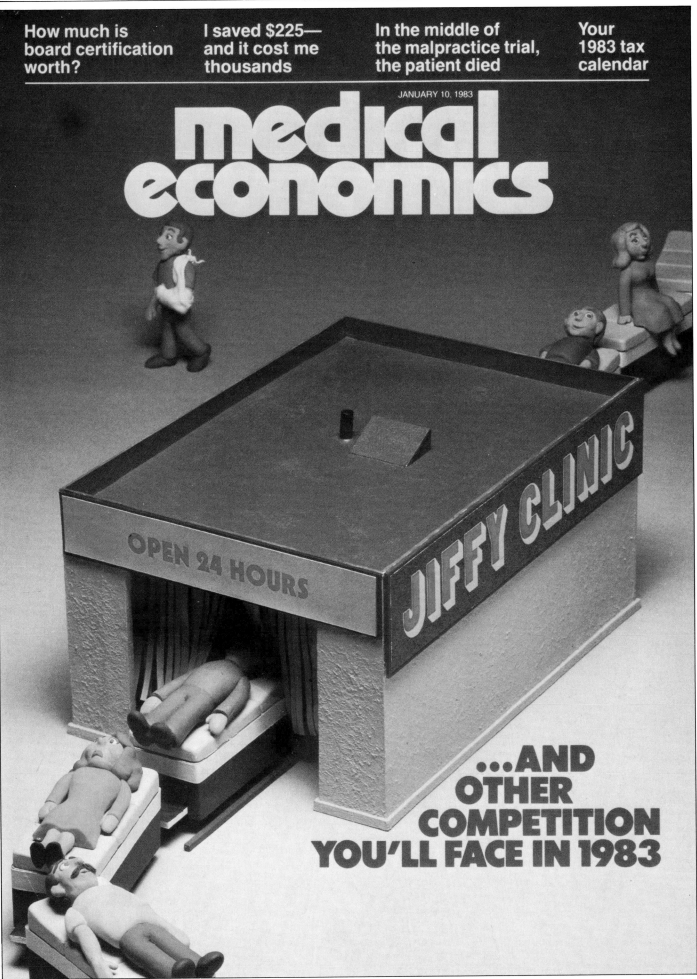

How much is board certification worth?

I saved $225— and it cost me thousands

In the middle of the malpractice trial, the patient died

Your 1983 tax calendar

JANUARY 10, 1983

medical economics

OPEN 24 HOURS

JIFFY CLINIC

...AND OTHER COMPETITION YOU'LL FACE IN 1983

Stephen E. Munz (art by Kathy Jeffers)

PROBLEM

The magazine had canvased senior medical students to find out whether this year's crop would distribute differently when they became licensed physicians—which specialties they might choose, whether they would prefer practicing near urban hospitals, and the like. The principal finding was that an increasing portion of graduating students said they wished to go into private practice and become office-based as quickly as possible. Since the magazine's main audience is composed of established doctors of this kind, this report had ominous competitive overtones. The editors, having decided to strum this string of concern, chose the working title: "Tomorrow's doctors: they want what you've got."

A *A newly hatched doctor breaks out of an egg. (How delivered)*

B *A new medical school graduate looks at the outside world with rainbow-filled binoculars. (Nature)*

C *An optimistic graduate is seen getting ready to bail out into his new profession. (Nature)*

D *The bright new crop of graduating doctors is portrayed with lightbulb heads. (Nature)*

A

B

C

D

SOLUTION

In this case an interpretive cover didn't seem the most effective way to reach the readers. The editors decided that a group of real senior medical students staring disconcertingly at our physician readers would bring the point home. The art director held casting sessions at nearby Cornell Medical Center and selected a fine group of highly credible models. (Nature)

The best
practice builders
cost nothing

What's behind
the fastest
money game of all

Help!
I'm a financial
disaster

Malpractice premiums:
hefty today,
huge tomorrow

NOVEMBER 28. 1983

medical economics

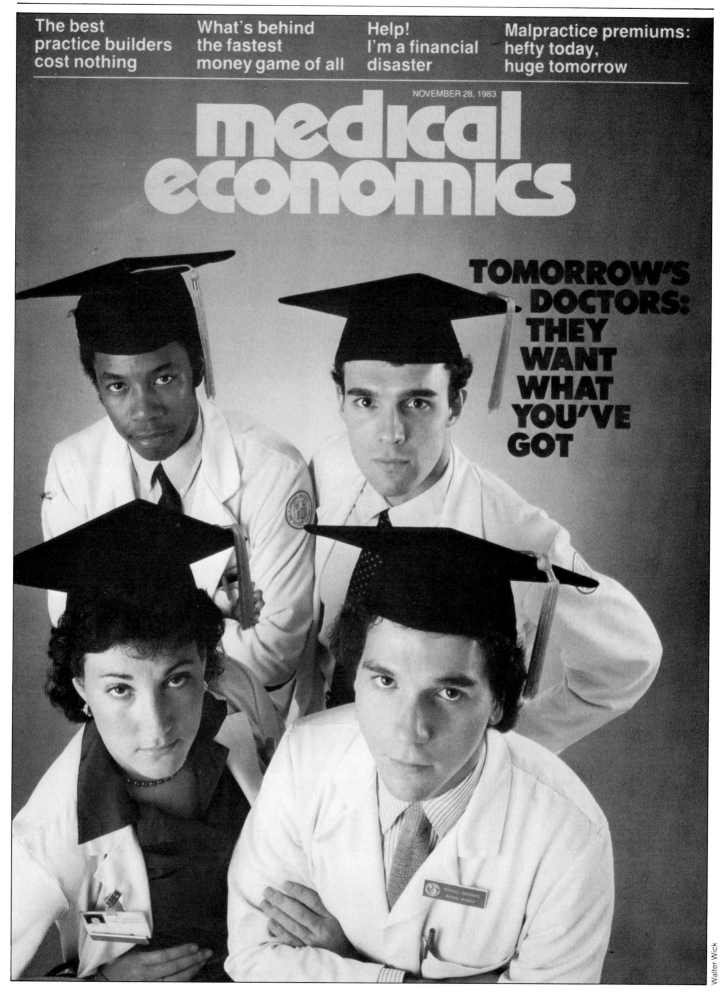

TOMORROW'S
DOCTORS:
THEY
WANT
WHAT
YOU'VE
GOT

Walter Wick

PROBLEM

A new method of paying for medical costs had become increasingly popular in many parts of the country through the operation of preferred provider organizations (PPOs). Here's how they work: A large corporation will tell its employees to use certain "preferred" doctors for all their health care needs, in exchange for which the company insurance pays most or all of the cost. The company will have shopped for the best rates in the community and, since flexibility of service is also needed, the companies usually will have chosen groups of associated doctors. This puts the lone practitioner at a disadvantage; he or she can't compete for the companies' benediction, and the individual practice may suffer. The working title was: "Can you compete with the PPOs?"

A *A lone doctor is shown facing an opposing line of PPO players in a football scene. (Nature)*

B *The lone physician is seen missing the wagon of opportunity. (Nature)*

C *White hats and black hats distinguish good guys and bad guys. Who wore which would depend on whether the editors were taking the viewpoint of the companies or the individual doctor. (Nature)*

A

B

C

SOLUTION

The final image shows the lone doctor left out in the cold; the title was adjusted to fit this image. (Nature)

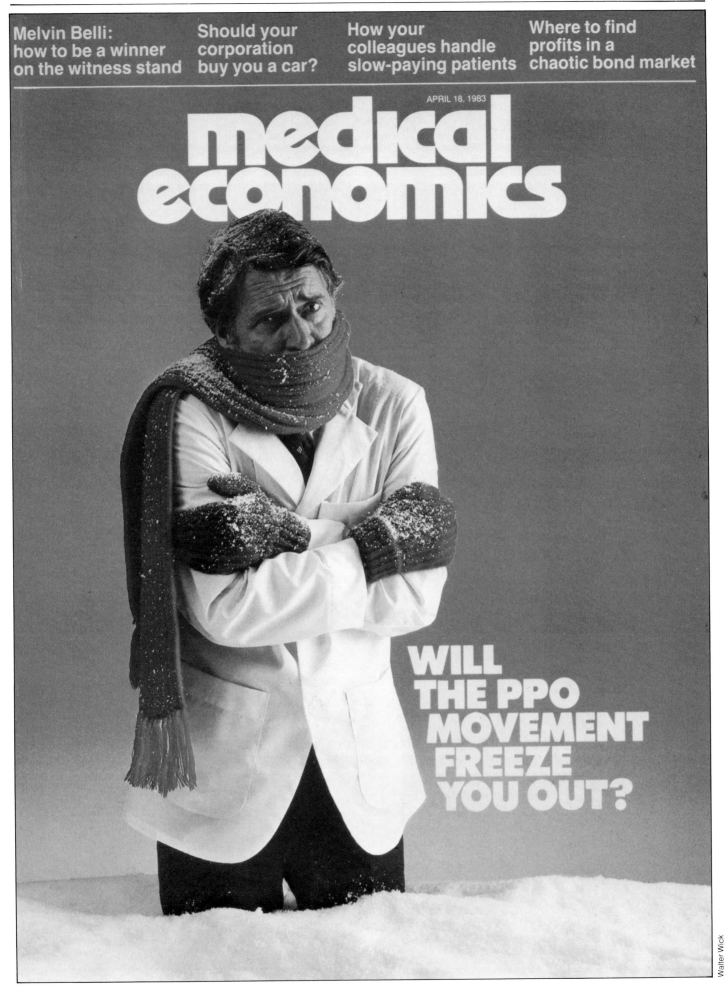

Melvin Belli:
how to be a winner
on the witness stand

Should your
corporation
buy you a car?

How your
colleagues handle
slow-paying patients

Where to find
profits in a
chaotic bond market

APRIL 18, 1983

medical economics

WILL THE PPO MOVEMENT FREEZE YOU OUT?

PROBLEM

Once every few years the magazine offers planning advice for readers who invest in common stocks. The working title for this story was "How to pick a long-term winning stock."

A *A small doctor is shown watering sunflowerlike company stocks. (Nature)*
B *The doctor's investment program is shown as a money tree. (Nature)*
C *The magazine's advice is represented as being on target. (How delivered)*
D *The doctor's share of investing profit becomes a pie. (Why given)*
E *The successful doctor investor rides a bull (market). (Why given)*
F *The doctor rides (triumphs over) a bear (market). (Why given)*
G *The investor/reader picks a winning horse. (Nature)*

A

B

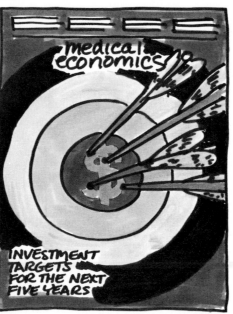

C

D

E

F

G

SOLUTION

The old story of the tortoise and the hare gave the cover its final image. The fun of executing this idea came largely from using live models. The photographer bought a baby cottontail a week before the shooting so that it would become accustomed to handling and bright lights, and the tortoise was rented from a pet store. The teenaged rabbit and the middle-aged tortoise became earnest colleagues and fast friends. (Nature)

Little errors that mean big malpractice trouble

Fringe benefits: frosting the corporate cake

A 30-minute checkup that keeps Medicare in line

"I have a right to threaten your life, Doctor"

DECEMBER 21, 1981

medical economics

LONG-TERM WINNERS AND LOSERS

Our special poll of investment experts

Walter Wick

PROBLEM

In these days of rising health care costs, many patients are choosing a type of treatment that saves money: same-day surgery. A surprising number of operations can be performed so that the patient is admitted and released in one day. Since the patient doesn't incur the expense of an overnight stay in the hospital, the savings are substantial and recovery at home is adequate. The working title was "Same-day surgery: why everyone is learning to love it."

A *A fully dressed, busy commuter is sitting on the operating table. He's looking at his watch to assure himself that the procedure is moving along well. (Nature)*

B *Our commuter businessman has had his operation and is wheeling himself happily out of the operating room with his briefcase on his lap. He's ready to resume work. (Nature)*

A

B

SOLUTION

The final idea was sparked by a casually dropped remark in the art conference, something about having to find a place to park at the hospital. Having to run outside to put another coin in a meter every hour is the ultimate in short-range planning. The appeal was heightened by having the patient wear his hospital examination gown while dropping in the coin. (Nature)

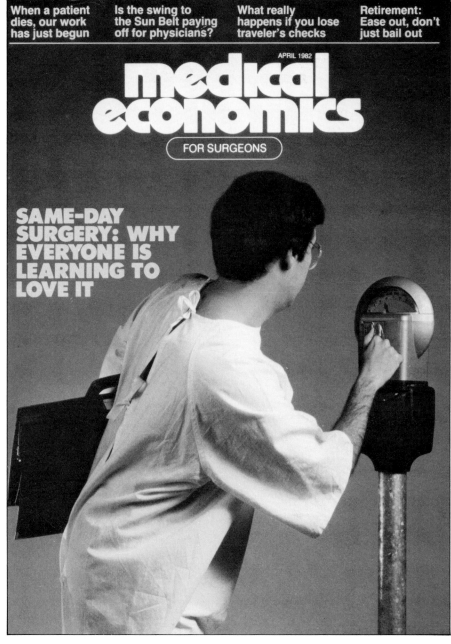

When a patient dies, our work has just begun

Is the swing to the Sun Belt paying off for physicians?

What really happens if you lose traveler's checks

Retirement: Ease out, don't just bail out

APRIL 1982

medical economics

FOR SURGEONS

SAME-DAY SURGERY: WHY EVERYONE IS LEARNING TO LOVE IT

Stephen E. Munz

PROBLEM

As the medical profession gets more and more crowded, the pinch of increased competition is particularly acute among surgeons. Some of the gentlemanly professional customs of yesteryear are giving way to more aggressive policies designed to get new patients and keep old ones. The working title for this was "Competition among surgeons takes a tough new turn."

A *Two surgeons are using a patient for rope in a tug-of-war. (Nature)*

B *Two surgeons are fencing. Presumably the winner gets the patient. (Nature)*

C *A challenge is delivered in the form of a quivering scalpel stuck between the fingers of a surgeon's hand reaching forward. (How delivered)*

D *One surgeon forbids another to cross the line. This is his professional turf. (How delivered)*

E *The opposing doctors are bracing for a Western-style showdown. (How delivered)*

F *The apprehensive doctor is shielding his patients from a raid. (Why given)*

G *This surgeon is really ready for trouble. He mans a 50-caliber machine gun from behind piled sandbags. (Source)*

A

B

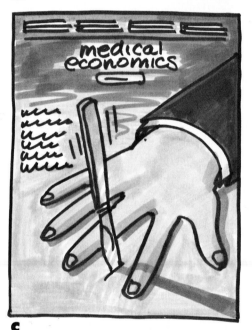

C

D

E

F

G

SOLUTION

The idea of competition as war seemed appropriate; the surgeon seeking to protect his practice is shown laying barbed wire and stacking sandbags around his base of operations, the surgical suite. (Nature)

Financial strategies that counter the latest tax rules

Testing Mercedes' "Baby Benz": Wow!

Congress draws a fresh bead on your fees

Melvin Belli: The new malpractice risks you'll face

AUGUST 1983

medical economics

FOR SURGEONS

OPERATING ROOM

COMPETITION AMONG SURGEONS TAKES A TOUGH NEW TURN

Stephen E. Munz

PROBLEM

A panel of experts convened by the magazine asserted that doctoring would be so different by 1990 that present practitioners would be numbed by the shock if they could be given a glance of the future. Increased government intervention and regulation, growing patient demands, and the greater need to advertise or market medical services would all conspire to make the doctor ask himself "Will private practice be worth the hassle in 1990?"

A *The private practitioner is shown mounted as a specimen of an extinct species in a glass case being ogled by the public. (Nature)*

B *The doctor is shown as an assault victim, bandaged after the battering he gets in 1990. (How delivered)*

C *The doctor is portrayed as Gulliver being tied down by Lilliputian bureaucrats and other pests. (How delivered)*

A

B

C

SOLUTION

None of the ideas touched enough bases or enough different problems that doctors would face in the next decade. The image finally chosen depicted an entire medical landscape in which the doctor, following a winding path that leads up and down hills, encounters the elements—Uncle Sam, patients in beds, hospitals, and other agencies—with which he must deal. (Nature)

How medical families spend their money

We had a warranty on our home and lost $40,000

When it comes to making rounds, later is better

Don't miss any of the new capital gains breaks

FEBRUARY 1, 1982

medical economics

1990: WILL PRIVATE PRACTICE BE WORTH THE HASSLE?

Our special section offers surprising—and often upbeat—forecasts about competition, patients, hospitals, outside intervention, and your finances.

Lonni Sue Johnson

PROBLEM

Peer review is an invaluable tool for physicians, who are responsible for making sure the quality of hospital care remains high. This tradition allows doctors to make charges of incompetence and other complaints against colleagues, and it allows the accused doctor to respond. The hearings are usually conducted by the medical board of the hospital and the proceedings are confidential. This confidentiality is not airtight, however, and testimony may leak, thus causing personal difficulty for a witness who has delivered a candid opinion about a colleague's performance. The story asks "Can you really speak your mind in peer review?"

A *The reputation-ruining potential of the hearing is shown by the soiled paper-doll doctor in a chain of others that are clean. (Nature)*

B *More damage to the testifying doctor is symbolized by thrown mud. (Nature)*

C *The witness doctor is shown being mangled by the sausage grinder of loose talk. (How delivered)*

D *The hearing board is represented by one doctor whose careless talk takes the form of daggers that wound the testifying doctor. (Why given)*

E *The testifying doctor is shown trying to speak in the hearing; his mouth is taped shut. (How delivered)*

F *The walls of the hearing room have ears. (Nature)*

A

B

C

D

E

F

SOLUTION

It was decided that the best way to represent the testifying physician's anxiety about speaking freely in the hearing was to show him imagining that the microphone was becoming a snake, that it was the focus of a threat he was feeling. A talented prop artist made a clay cobra, painted it realistically, added a disturbing little vinyl tongue, and mounted the sculpture on a microphone stand. Result: one genuinely jittery doctor. (How delivered)

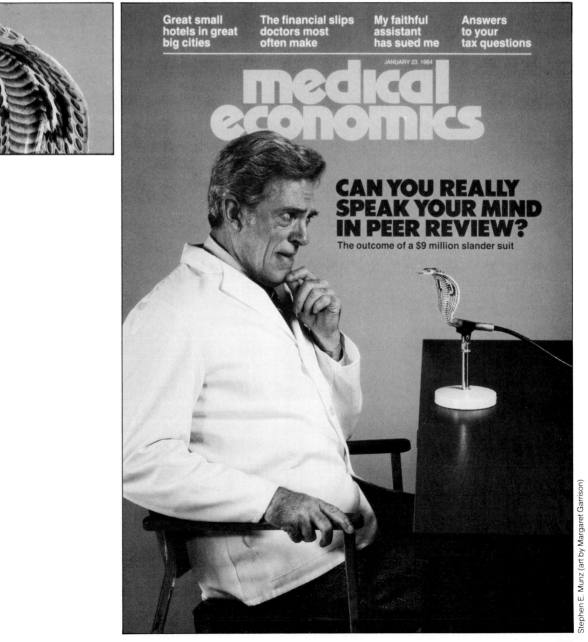

Great small hotels in great big cities

The financial slips doctors most often make

My faithful assistant has sued me

Answers to your tax questions

JANUARY 23, 1984

medical economics

CAN YOU REALLY SPEAK YOUR MIND IN PEER REVIEW?

The outcome of a $9 million slander suit

Stephen E. Munz (art by Margaret Garrison)

PROBLEM

The magazine acquired an unusually moving and well-written story by a west coast doctor whose daughter got involved heavily with drugs. The agony of the family and the long, tough path to rehabilitation made this a valuable article for readers both as physicians and as parents. The working title was "What we can all learn from my daughter's battle with cocaine."

A *The girl is shown hitting bottom; she's slumped listlessly in an armchair. This would have been portrayed in a moody photograph in which the face and upper body were darkened. (Nature)*

B *The daughter's picture and drug paraphernalia are used to make a poignant still life. (Nature)*

C *Cocaine—baby powder, that is—is used to draw a skull and crossbones, the symbol of danger. (Nature)*

D *The danger is represented by a snake poking through a pile of coke. (Nature)*

SOLUTION

The girl and the cocaine became one; her portrait was drawn on a plate of glass, and the red background paper was separately lit so that the face acquired a glow and a floating quality. (The art director was told that if the powder used in the picture were real cocaine the street value of the amount shown would have been around $10,000.) (Nature)

DECEMBER 12, 1983

medical economics

WHAT WE CAN ALL LEARN FROM MY DAUGHTER'S BATTLE WITH COCAINE

Stephen E. Munz

PROBLEM

Government intrusion is one of the biggest bugbears for all kinds of physicians. Government-imposed fee controls and ever more complex practice regulations often make the average doctor feel as though he or she is being suffocated. Such federal strictures make it ever tougher for doctors to practice in their field of training. The working title was "Will Washington keep its promise to leave doctors alone?"

A *The physician is trapped under a large federal hat. (Size)*

B *Uncle Sam's hand is squeezing the doctor. (Size)*

C *Once again, Uncle Sam is on the doctor's back—literally. (Weight)*

D *The government, represented by the Capitol Building, is burdening the doctor. (Weight)*

A

B

C

D

SOLUTION

It was decided to use the image of a huge Uncle Sam peering ominously over a hill. A late modification eliminated the foreground figure of the doctor so that Uncle Sam would seem to be menacing the reader; the headline was altered to reflect this, too. The photograph is a one-piece shot made with a miniature landscape. (Size)

How private eyes expose phony malpractice claims

Now, more than ever, you're a target for bad advice

Will M.D.s win a counterattack on chiropractors?

Forms that make clinical paperwork almost painless

JULY 19, 1982

medical economics

WILL WASHINGTON KEEP ITS PROMISE TO LEAVE YOU ALONE?

PROBLEM

The maiden issue of a special edition of Medical Economics *for surgeons appeared in March 1982. It was felt that a major survey of average incomes of several kinds of surgeons across the country would give the magazine a strong start. The working title was "Surgeons' earnings: where do you fit in?"*

A *Earning power is represented by a heavily muscled surgeon's arm holding a scalpel. (Nature)*

B *The high-paid surgeon could be cast as Superman. (How delivered)*

C *The ways to high and mediocre earnings are shown as a fork in the road, with an appropriate sign for each path. (How delivered)*

D *A high earner could be symbolized by a surgeon's foot clamped in a rollerskate. (How delivered)*

E *The busy surgeon could be shown actually living in the operating room with easychair, slippers, newspaper, and other accessories. (How delivered)*

F *The secret of the high-earning surgeon's productivity is revealed — six arms. (How delivered)*

SOLUTION

A minor change produced the final image. The surgeon is preparing to go to work, but he's looking at the gloves instead of toward the title type that talks about money. This sideward-looking pose and the headline had a strange effect; the doctor seemed to be eager to get to work to earn money. Having the model look downward avoided this implication; the result was an image of the confident, competent healer ready for action. (Nature)

Stephen E. Munz

PROBLEM

Malpractice insurance premiums had reached incredible levels mainly because the courts were awarding damage payments as high as seven digits (one of the largest court awards to date was a five million dollar decision handed down in Florida). The magazine discovered that the insurance companies intended to raise the premium rates again (basic protection ran to $20,000 or more per year before then), and wanted to prepare readers for the new hike. There was no definite working title on this story—just the subject of rising rates and the shock this would cause among doctors who had to have this protection.

A *A doctor is shown on a treadmill. Rising insurance rates make him feel he's standing still economically. (Nature)*

B *The surgeon prepares to fight back. (Why given)*

C *Doctor-written graffiti on the wall explores the problem. (Why given)*

A

B

C

SOLUTION

The editors could imagine that the average surgeon's reaction might be extreme, that he or she might do something desperate. The image of a doctor preparing to jump from a building upon hearing the bad news seemed to be the most attention-getting way to express this shock. This is actually two photographs stripped together; the photographer shot his own feet on the second floor ledge and then took the top picture from the seventeenth floor. (Why given)

Building a staff
that puts your
best foot forward

How to soar
when the market
takes off

Get just the
car you want
at your price

Who closed
the hospital door
to this doctor?

MAY 1982

medical economics

FOR SURGEONS

DON'T PANIC WHEN MALPRACTICE RATES GO UP AGAIN

Walter Wick

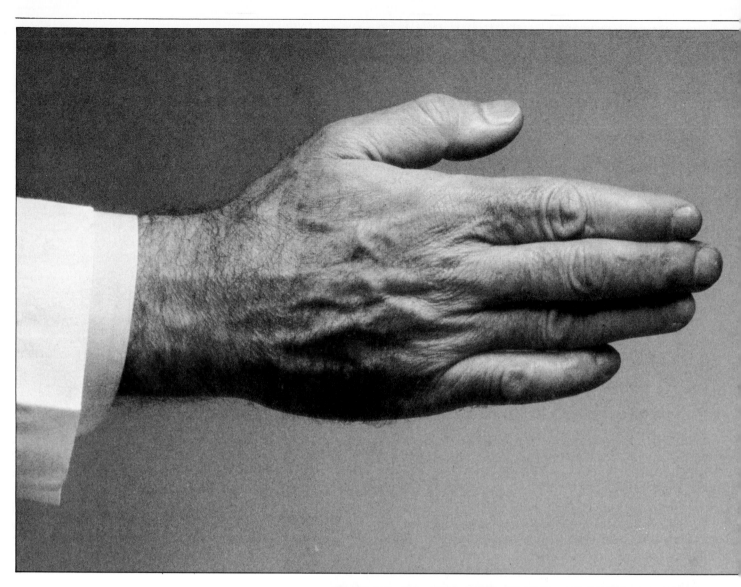

SHAPING YOUR CONVERSATION WITH THE READER

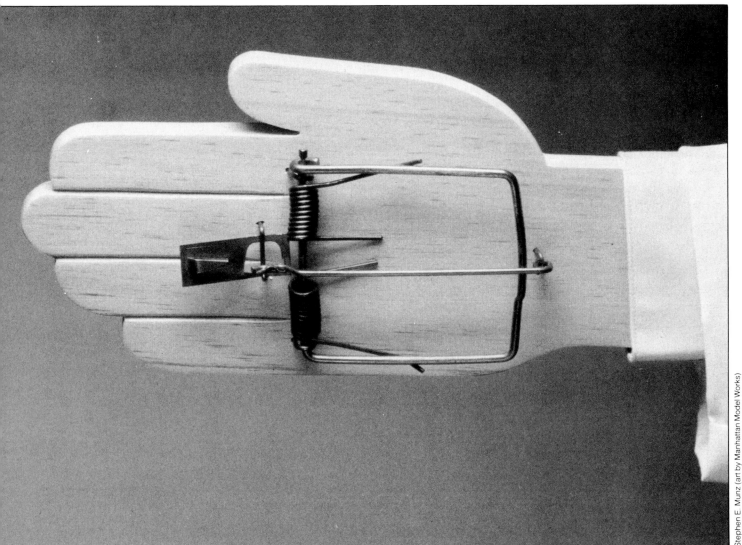

This image was used to illustrate a fraud story in Medical Economics. *The title was "A handshake cost me $15,000."*

In the simplest terms your job is to deliver a message from the author to the reader. If you make an ad, create a poster, or execute a piece of promotion mail, the message you deliver is comparatively short. If you make a television commercial to convey a thought, the dialogue will be somewhat longer; it may even be delivered by a character who talks directly to the viewer.

However, if your business is making magazines, your relationship with the reader is longer and more complex. For one thing, you are attempting to build a lasting association with your audience; the reader should sense a friend in the pages from issue to issue, a voice he can trust and a person whose company he enjoys. If your publication is to do this—to create its own voice— some thought must be given to overall pacing.

Of the different types of communication treated in this book, the magazine most resembles a spoken conversation. Think of the most satisfying conversations you've ever had. In most cases, the person with whom you were talking probably had a great deal to do with the vividness of your memory. He or she was articulate, chose colorful and appropriate terms to describe thoughts, was animated, and demonstrated enthusiasm for the main points of a presentation. And, last but not least, he or she put variety in the delivery.

So, too, a magazine should put variety in its pages—a few well-distributed peaks of graphic excitement interspersed with valleys of quieter material. Just as a skilled conversationalist will raise and lower his or her voice for effect, so a magazine should alternate the strongest and most emotional graphic blasts with stretches of expository chatter. The quieter parts of the presentation give dimension to the flamboyant highs as the reader turns the pages, lured from one section to the next by the appearance of something momentarily different. Just as the listener would lose interest in a speaker who droned through a lecture with his or her voice kept at the same level, the reader will be bored and left unstimulated by a magazine that is undifferentiated throughout its length, that either bumbles along with a flat, low-keyed mixture that lacks dramatic highs or, on the other hand, tries to maintain a constant high-decibel level with an endless, unrelieved string of fireworks. Both structures can be dull.

The ideal distribution of graphic excitement in a magazine can be shown as a series of dots drawn along a straight line. The dots represent elements within the body of the magazine—a feature lead illustration, a stretch of type without graphics, or a chart group of visuals—and the size of the dot varies according to the visual wallop the item delivers. A huge picture spread showing a closeup view of John Belushi's face scowling directly at the reader would merit one of the biggest dots on the line of elements in that issue of *Rolling Stone*. Four pages of text with a few scattered display quotes in the same issue, however, would show as very small dots—perhaps a sixth the size of the picture-spread dot. This control device, the linear mapping of the emotional profile of a magazine, can be a valuable tool. It allows the designer to pace an issue to avoid both problems: long flat areas without graphic spice, and graphic blockbusters that are bunched in one part of the magazine at the expense of other parts.

So it becomes apparent that excitement, graphic drama, must be spread throughout your magazine if it is to attain variety and interest through change of graphic pace. Let's concentrate on the emotional high points, the peaks of excitement in the body of a magazine. There are twelve main tools a designer may use to create maximum surprise. They are:

1. Bright color
2. Mood photography or illustration
3. Action
4. Sudden change of scale
5. Large type
6. Subject focus on the reader
7. Unusual point of view
8. Emphasis on telling detail
9. Unorthodox layout shapes
10. Repetition
11. Unexpected elements or dimensional mix
12. Visual puns

An illustration or photograph may use one or several of these factors, but these are the basic ingredients for getting the reader's attention. The twelfth tool, visual puns, is the strongest weapon in the graphic arsenal, the one most satisfying to use. The preceding chapter dealt with the critical function of getting ideas, of generating striking, memorable images to lend impact to the writer's message. The Bite System is, primarily, an approach to making visual puns, symbols, graphic analogies, and curiosity-provoking images to dramatize words. Conceptual surprise of this sort is a very powerful design device and it comes to mind readily. But the remaining eleven graphic tools deserve some attention, too.

1. Bright color. More than enough has been written about the psychological impact of color—the emotional effects specific colors can have—for me to add much of use. The color used, its hue and intensity, have a marked mood-affecting capability: blues and greens are soothing; reds, magentas, and oranges are stimulating; and browns and grays are conducive to quiet reflection.

However, the sense in which bright color can be effective as a graphic tool is its unexpected use in large amounts; the closer it can be to a primary hue—red, yellow, or blue—the more powerful its impact will be. Color can pervade an entire illustration or may be used in select parts of the visual; often it can be used to emphasize a function within a diagram or photographic demonstration. If you apply bright color as a graphic device, it is important that you not be timid about using it, that you not be cautious about the amount or brilliance of color you use. Of course, some care must be taken to allow any surprinted type to read easily; small black type in a caption, for instance, can be very difficult to see against a strong color other than bright yellow. Bright color is a potent graphic weapon that you should stay alert for opportunities to use.

2. Mood photography or illustration. Few devices do more to engage the reader than the use of a large photograph or illustration that is rich in emotion or that displays a strong feeling for a place or situation.

THE EMOTIONAL PROFILE OF THREE MAGAZINES

Every publication develops its own personality; it speaks in a unique way and with a distinctive tone of voice to its readers. One of the best ways to understand and control this voice is to diagram the graphic highs and lows in a given issue, as shown on the opposite page. The reader's progress through the book is represented as a line on which different sized dots, indicating large or small graphic elements, have been placed. The vertical lines define the main feature well, the first group of major articles. How each of these magazines distributes its excitement can easily be seen.

Most of the biggest blasts are grouped in the first part of the feature well but each department carries a larger-than-normal graphic display.

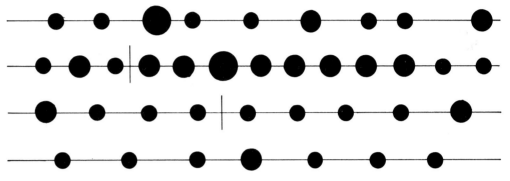

This book distributes its graphic fireworks fairly evenly throughout the issue. There's always something bubbling for the reader.

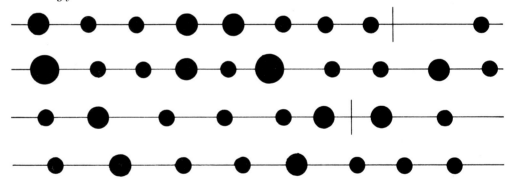

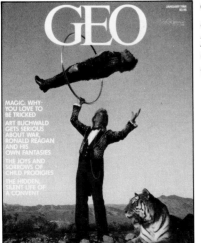

One of the best remaining markets for spectacular photography, Geo produces a continuous cannonade of exciting pictures for most of the middle of the book.

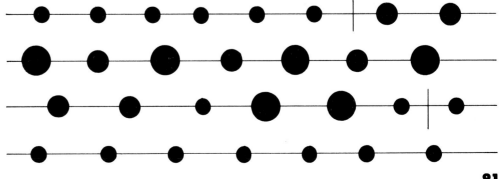

Such a picture can invite the reader into the page, can make him feel a part of what is happening in the picture. A brooding panorama, a dramatic piece of reporting, a tableau of tension and danger—these are effective mood statements.

3. Action. Readers, people in general, react to pictures of objects in motion—the blurred form of a polo pony streaking toward the goal, a leaping dog frozen in midair as it reaches for a frisbee disk, a mountain climber whose rope has snapped and lets him begin his long fall. The action may be happening or obviously about to happen; something is in a state of serious imbalance, teetering precariously on the edge of disaster. The painter Willem de Kooning once referred to hunting for the "state of becoming" pose for his subject: the person may be shown just as he is uncrossing his legs and is in transition from one static pose to another. De Kooning prefers to catch his subject in action even if the action is small; he likes the position in which something is changing to become something else. This is another way of saying that pictures in which people or things are shifting are generally more interesting to the reader than scenes in which everything is settled, everything resolved. There are motionless subjects—groups of people, for example—that imply movement in the immediate future; the group may display so much tension in the individuals' body positions or expressions that they are clearly ready to explode into action. This is implied action; its use would be comparatively rare. Overt action in an image is the form this graphic tool usually takes.

There was a time when no magazine or book editor would use a "fuzzy" photograph, that is, one in which every part of the subject was not in clear focus and efficiently frozen. In the last couple of decades, however, most communicators have learned to appreciate the special qualities of the intelligent use of blurred action shots. The image of a deer, for instance, may be clear except that its legs are lost in swipes of color, and the entire background of the picture simply shows as blended bars of moving detail. The combination of clarity and definition obtained by panning the camera—keeping the camera trained on a subject that is moving across the photographer's field of vision—can make the page seem much larger than it is. The action implies that much more is happening outside the bleed area of the page.

4. Sudden change of scale. Few things will be more surprising to the reader than to encounter a larger-than-life photograph that fills the page or spread. An inexperienced designer will often produce several pages of pictures that show the subjects all at roughly the same scale. This can occur when the designer has too many photographs for the space and feels compelled to use them all. It is far better to cut the selection and better yet to go to extremes by enlarging and cropping one or two. Crowd the limits of the page with a shot; if it is a human face, enlarge until all that show are eyes, nose, and mouth—and enlarge even more if the eyes tell the story by themselves! Change of scale is the graphic equivalent of the change to fortissimo in music, the unexpected fanfare of trumpets after the quiet and melodious passage in a symphony.

5. Large type. This tool creates the same kind of surprise as sudden change of scale: the reader turns the page to confront one or more words that are too large to ignore. The most important requirement for the use of superlarge headline type is that the words you want to emphasize be significant to the reader. If they are not, the effect will be more comical than useful. A headline I remember seeing gave great force to the first two words and let the others trail off into insignificance; the title read, "THE MAN who knew enough." Who cares about THE MAN? The term had been artificially separated

(Text continued on page 162)

THE TWELVE BASIC TOOLS FOR CREATING VISUAL SURPRISE

The graphic designer, like an artist of any type, has a surprisingly simple palette. Just as a musician can create endlessly different moods with a few standard notes, the magazine designer can orchestrate an issue with excitement and uniqueness time and again through sensitive use of these twelve basic devices. Of course, two or three of the tools may contribute to the power of a single piece, but normally, each carries enough impact to do its part in shaping the issue.

Bright color

Large amounts of color can evoke a powerful emotional response from the reader. The more intense the color and the more primary the hue— pure red, yellow, or blue—the better. The illustration on the facing page, the surgeon's guide to informed consent, shows bold use of color.

Andrea Baruffi

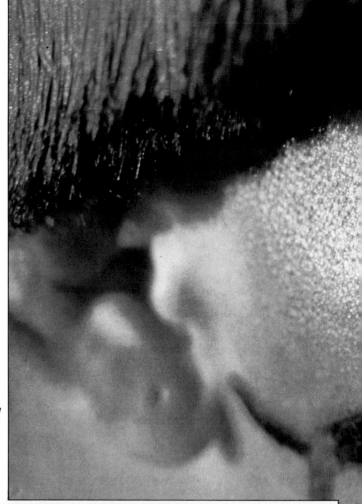

These two spreads from Geo *magazine demonstrate the ample shock value of large amounts of bright color. At right is a close look at a young man from the Xingu tribe of Brazil. Below is the first in a series of portraits of rich young video game designers.*

GEO**PEOPLE**

JOYSTICKS AND ROYALTIES

A NEW BREED OF YOUNG MILLIONAIRE IS AFOOT IN THE LAND: THE DESIGNERS OF VIDEO GAMES

Joel Gluck, a sophomore at MIT, sold his first video game when he was 16. Three years later, he is still collecting royalties.

72 GEO

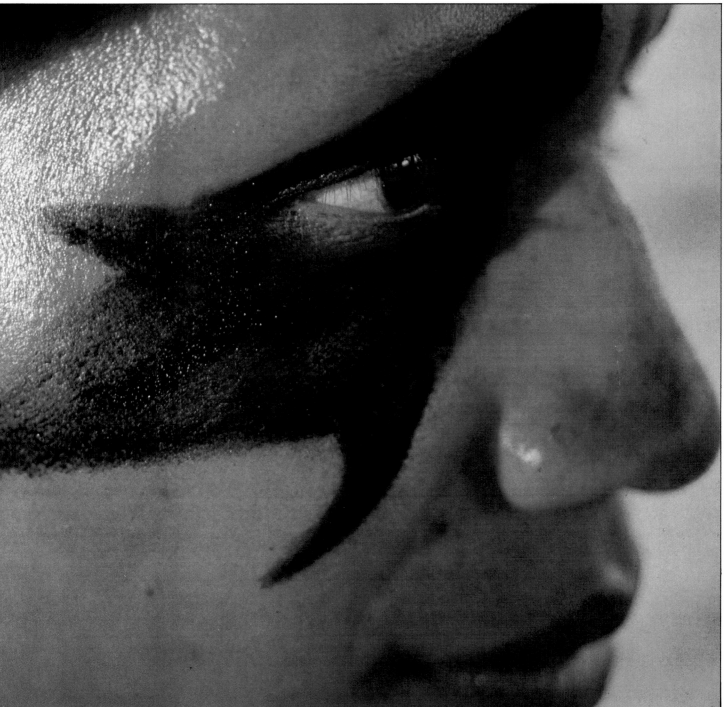

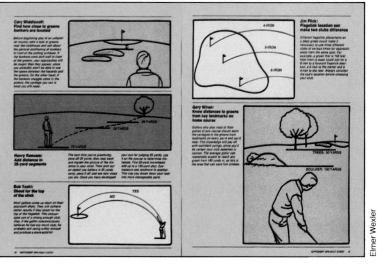

The most visually mundane material can become exciting and demand attention through use of color. This is a series of tips from Golf Digest's *playing editors — advice about estimating distance on a golf course.*

Elmer Wexler

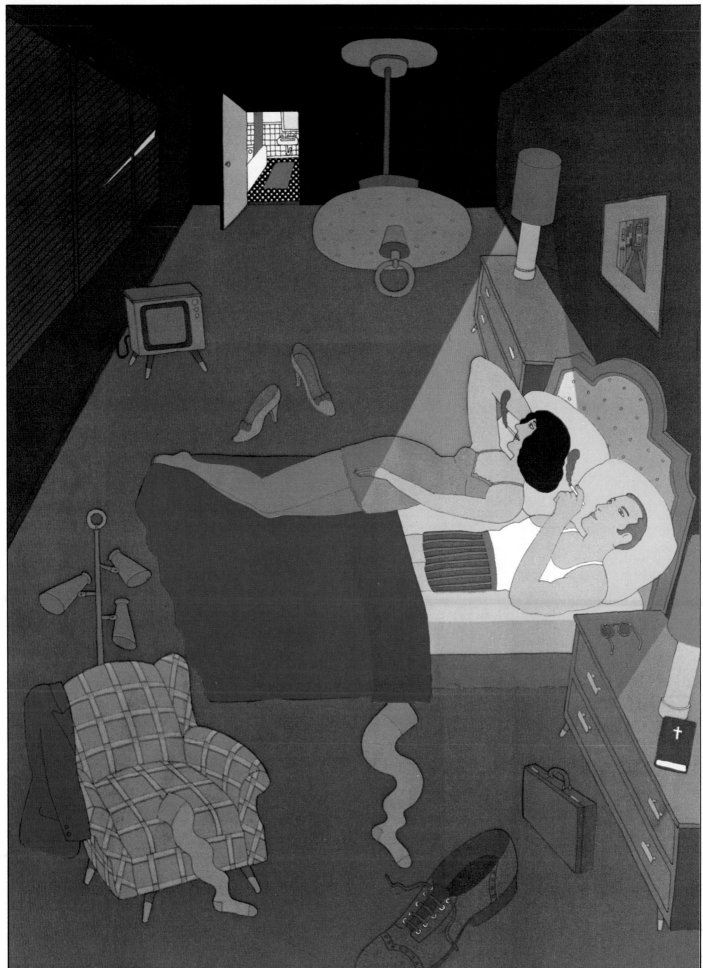

Seymour Chwast

Seymour Chwast's richly colored interpretation of a tryst in a hotel room shows more subtle use of color in the bright islands of hue against a dense background. This strong statement uses color to set a specific mood.

This symbolic statement about a court defeat of California bureaucrats was given extra clout through the liberal use of strong color.

Stephen E. Munz (art by Kathy Jeffers)

The national mania for tennis was at its height when this spread was published. The explosion of color echoed the furor.

Mark Kozlowski

Mood photography or illustration

*Brooding landscapes or paintings that convey
powerful mood can be effective ways to get
attention. This tool is one of the most involving of
the twelve; the reader is often drawn into the scene.
It can allow the viewer to feel that he or she is
there facing seldom-seen Spanish castles or being
suspended above unclimbable mountain peaks.*

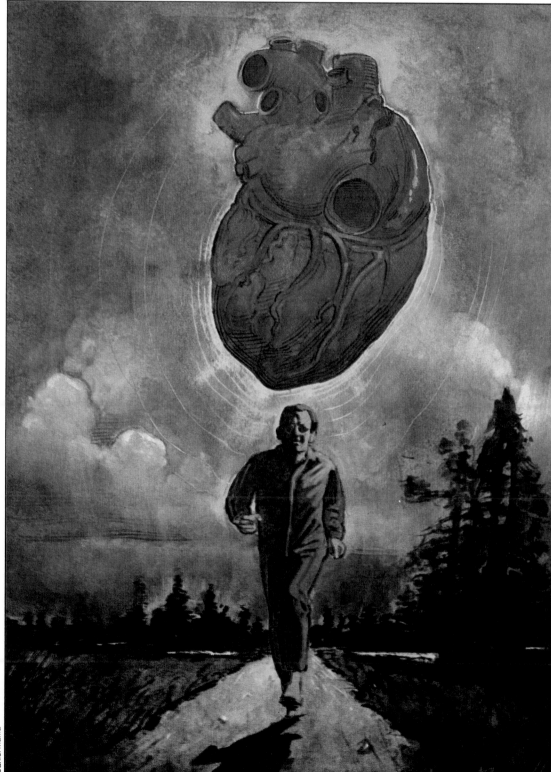

Daniel Maffia

*The article is about a
doctor who brought on
his heart attack by
overtraining. The heart
looms above him as
he runs to get fit.*

CASTLES IN SPAIN

Menacing, outrageous, monstrous, barbaric — the Spanish speak of their legendary castles in the most unflattering ways. But the demonic beauties are irresistible.

Reinhart Wolf © 1983 Knapp Communications Corporation

Reinhart Wolf made a series of classic portraits of forbidding Spanish castles for Geo *magazine. This one is the fortress of Coca in Segovia, a castle noted for its sumptuous and decorative brickwork.*

John Newcomb

The last few seconds of sunlight cast a glow over the Firestone North golf course in Akron, Ohio. The punishing nature of the course is revealed in a glance.

Galen Rowell

Galen Rowell

The incredible story of one of the most spectacular motion pictures of our time

A BRIDGE TOO FAR

MOVIE SPECTACULAR VOL. 1, NO. 1

Robert Penn

The title spread of a newsstand book created to publicize the film of the same name shows one of the key scenes from the battle, the burning of the Arnhem Bridge, an Allied objective in Holland.

FINANCIAL PORTRAIT

"WE'RE LAND-RICH AND CASH-POOR"

There's not much spendable income, but a rugged country lifestyle of hard work and hard play keeps this GP and his family happy.
By Robert Cassidy MIDWEST EDITOR

At sunrise in Readlyn, Iowa, the air is so crisp you can almost taste it. Huddled in the duck blind, shotguns at the ready, are GP Clinton E. Berryhill and his 14-year-old son Todd. A lone duck flies in low from the west, barely 15 yards from the blind. Berryhill stands, raises his 20-year-old Sears 12-gauge and fires. He wings the duck, and Todd quickly finishes off the bird with a couple of shots from his 20-gauge pump. The small teal will be all they bag this morning, but for Berryhill, 54, the hunt is an invigorating way to start the day. And his days are long ones.

With patients scattered throughout a 30-mile radius of this rich corn, soybean, and hog-farming community in northeastern Iowa, he's on call 24 hours a day, 365 days a year. He's also the town pharmacist, dispensing $50,000 worth of drugs a year. He's even an assistant medical examiner in his spare time. An exhausting regimen? "I wouldn't have it any other way," Berryhill answers. "I work a lot and play a lot, and most of the time I can't tell the difference."

As for his personal life, Berryhill says he has everything he wants: four grown children by his first wife, who died in 1965, and three youngsters by Linda, his 37-year-old second wife (and office nurse), whom he married in 1966. He also has 200 acres of cornfields, hog pens, ponds, and timberland plus a sideline business—an 80-acre sand-and-gravel quarry. "My material needs are extremely few," he says. "The number of things I don't give a damn about would fill a very large book."

Beneath this lighthearted exterior, though, there's considerable disquiet. "My financial life is in chaos," Berryhill frets. As the accompanying box on page 95 details, Berryhill is at the crucial point in a series of real estate ventures that have made him "land-rich and cash-poor." For the last two years, he's had to devote most of his net practice income to paying off his loans. That cost him more than $70,000 of his 1981 practice net of $106,300. If he can come up with another $75,000 or so this year, he'll wipe out his real estate debts and end up with a net worth of more than $1 million.

Living on a spendable income of

MEDICAL ECONOMICS/APRIL 12, 1982 **91**

Everything the Berryhills want in life is close to home. One favorite spot: the fish pond on their 200-acre farm, where oldest son Kenneth and his wife, Jenny, savor the tranquility.

Winfred Meyer

The allure of inaccessible places is the theme of the two pictures opposite. The top of Peak K2 Karakoram sports a cloud cap, while the desolate edge of Patriarch Grove in California displays the stubborn shoots of age-old live oaks.

Tranquility radiates from this scene of a family enjoying a quiet evening of fishing on their own pond.

Action

Violent movement in a large picture can be a compelling device. People are attracted to situations where things are happening, objects move, and conditions shift. As in life, where a street crowd quickly collects around a group of breakdancers or some other corner performance, so it is with print—an action photograph or illustration can attract readers.

Another scene from the realistically filmed war story, A Bridge Too Far, *shows soldiers being blown in the air by artillery shells.*

Dennis Chalkin (acrobat: Michael Heintz)

Action is a mild term to use for this domestic scene. The staircase is only a façade and the flying father is Michael Heintz, an acrobatic dancer, shown here doing a forward flip from a platform behind the fake staircase.

A freewheeling hand-and-foot race at a Fourth of July festival offers a fine example of the use of pan-blur photography. By using a comparatively slow exposure and by moving the camera with the subject during exposure, the photographer did more to indicate motion than would have been possible with any number of "perfect" stop-action shots.

TES'

EVER
THEY
IS TR

Majestic pow
and legendar
retained in t
By Allyn Z.

85,000 Rolls-Royces
still on the road.

So you can imagin
to spend a week roa
Royce's first new m
scious, I settled beh
touched the gas pe
it—out of the garag
hattan's streets.

With only 338 mi
record 1 million, I a
to take liberties wit
6.7-liter V-8 isn't c
until it's ticked off
It didn't take me
Silver Spirit is not a

The Rolls Royce automobile is made for motion—what better way to show it? The car doesn't lose any of its beauty by being photographed in use.

Bob Hewitt and Frew McMillan are the premier doubles team in pro tennis. They are also complete opposites in most ways. This article answers the question:

WHAT MAKES THE ODD COUPLE THE WORLD'S BEST?

BY PETER BODO, Contributing Editor

Professional doubles exists in a twilight world of courts barely within shouting distance of the stadium—courts that are hard by dusty parking lots with depleted line crews and Spartan bleachers surrounded by crabgrass.

Matches are scheduled for convenience, rather than interest, at absurd times on short notice; it seems that during an outdoor tournament, somebody is always starting a doubles match when the sun is just a crimson rag on the horizon and most spectators are beginning to wish they had brought

along their sweaters. Indoors, it is axiomatic that at least one doubles match at any given tournament begins at midnight, when the sight of eager, fresh players warming up contrasts with the empty seats, paper cups littering the aisles, and the cloud of blue smoke clinging to the ceiling lights.

Doubles does not even attain the status of an opening act; it closes the show, when it can hardly be anything but anticlimactic. It is the poor relation of the singles game, dominated and overshadowed by the drama, glamour, tension and contained fury

Hewitt (left) and McMillan lunging at the net.

John Newcomb

Bob Hewitt and Frew McMillan dominated the men's doubles finals at the U.S. Open for several years in a row. This story lead catches two of their best moments of movement.

106

LS

ING

me comfort,
All are
lls-Royce.

EDITOR

to be said that
nly sound you
hear while riding
s-Royce was the
f its clock. Now
nce; the clock is

nprovement is in-
of the glacial
that have taken
his legendary ve-
e the first one was
ears ago. Now as
Rolls-Royce is the
k of automotive
d quality. Of the
ore than half are

hen I was invited
ilver Spirit, Rolls-
rs. A bit self-con-
and ever so gently
glide is more like
rly-burly of Man-

neter calibrated to
atiously, reluctant
ighty engine. The
roughly broken in

that the $109,000
car. It's a straight-

NOMICS/JAN. 18, 1982 **89**

Anthony Vaccaro

Change of scale

*Few graphic devices are more surprising than sudden,
extreme change in the distance between reader and
subject. The shock of going from conversation range
to eyeball-to-eyeball closeness is tremendous. This is
definitely one of the bass drums in the graphic orchestra.*

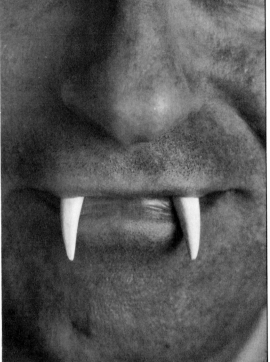

Stephen E. Munz (art by Asdur Takakjian)

WHEN A REAL ESTATE BROKER SMILES, LOOK FOR SHARP TEETH

It's simply good business to be friendly
with a buyer. But a broker works for the
seller—and he's out to earn his keep.

By Judith Trotsky SENIOR ASSOCIATE EDITOR

Several years ago, a Seattle OBG decided he could make some money if he invested in real estate. He is a man who invests with care, and his approach to this venture was no exception. He first educated himself in the techniques of finding and financing the right buildings, and then spent months deciding precisely what to buy. His guide through the lengthy search was the broker who handled the first building he went to see.

"He was *my* broker," the doctor says. "I told him what my goals were, and that I was a novice. He understood perfectly, asked me a lot of questions, and really wanted to help."

The doctor pauses. "At least that's what I thought. But after we'd finally closed on a complex of three small apartment buildings, I found out that he'd been trying to unload them for almost two years. I paid a lot more than what the owner would have settled for."

In fact, he paid $60,000 more than he had to for the properties.

He has an acceptable investment, but certainly not a great one. Careful though he was, the OBG had missed a point that's crucial to real estate dealing.

Whose side is a broker on?
A good salesman—any good salesman—has to be concerned with what the customer wants. But something is different in real estate. Buyers often forget who "their" broker really represents—the seller.

That doesn't mean the broker is out to cheat a buyer. An ethical broker knows he's not going to get a sale and the accompanying commission unless he can get the buyer and seller to agree. So he's got to get the seller to come down and the buyer to come up in price.

Price isn't the only item he may negotiate. John Belo, whose New York City firm, Kaplon-Belo Associates Inc., specializes in investment property, illustrates how a broker can work for both sides.

Belo's client, the seller, had a building for which he wanted $125,000. The price was high, but he refused to come down. The prospective buyer was about to walk away when Belo found a way to save the deal. He persuaded the seller to accept a mere $10,000 down and to take a mortgage at 8½ percent. "I got the seller his price and the buyer his terms," Belo says. And, not incidentally, earned his own fee.

That's the critical role you should expect the broker to play: mediator. "Many sellers complain

156 MEDICAL ECONOMICS/AUGUST 9, 1982

THE PMC STORY

We at Poongsan Metal Mfg. Co., Ltd. of Seoul, Korea, are proud to introduce our company to you, the ammunition dealers of America. Over the past three years, PMC has enjoyed remarkable growth in the ammunition market. Sales have tripled with each year and 1980 promises to be another record year. We take pride in a reputation built with uncompromising standards for producing the highest quality in munitions. Our success is due not only to the high quality of our product line, but also to your continued appreciation of our efforts.

PMC owns and operates one of the largest and most modern brass mills in Asia. Over 12,000 people are under our employ in four separate facilities. PMC currently occupies 1,600 acres of land and 3.5 million square feet of manufacturing space.

In North America, the Patton and Morgan Corporation is the exclusive importer of PMC ammunition. Our marketing strategy has been simple. We have worked in cooperation with regional master jobbers to offer the best product at a reasonable price. Sales figures for PMC munitions confirm the wisdom of this strategy since PMC master jobbers and their customers earn profits well above the industry norm. Because of the growing popularity of our products in this country, PMC is further expanding.

its product line. In addition to our regular military rifle ammunition, we now offer a wider selection of rifle and pistol cartridges.

If you would like more information on our products, please call us in New York at (212) 755-5530. Thank you for your enthusiastic reception of PMC ammunition.

Very truly yours,

Chan U. Ryu, President
Poongsan Metal Mfg. Co., Ltd.
Seoul, Korea

Changkyu Kim, President
Patton and Morgan Corporation
New York, N.Y. 10022

THE PMC PRODUCTS

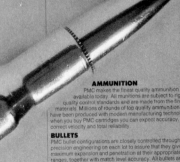

AMMUNITION
PMC makes the finest quality ammunition available today. All munitions are subject to rigid quality control standards and are made from the finest materials. Millions of rounds of top quality ammunition have been produced with modern manufacturing technology; when you buy PMC cartridges you can expect accuracy, correct velocity and total reliability.

BULLETS
PMC bullet configurations are closely controlled through precision engineering on each lot to assure that they give maximum expansion and penetration at their appropriate ranges, together with match level accuracy. All bullets are tested for the correct in-flight and on-target characteristics. Every bullet lot is subjected to stringent tests with the most modern equipment.

PRIMERS
PMC primers are designed for proper sensitivity and function. All boxer primers are non-corrosive and will not cause barrel rust. Close control of the diameter and depth of each primer during manufacturing assures the correct cartridge case fit.

POWDER
PMC smokeless powder is engineered to give each bullet correct velocity without flash or fouling. All powder is screened to prevent contamination and to insure consistent grain size. Every lot is checked for uniformity of burning rate.

CARTRIDGE CASES
Cases are manufactured in PMC's own brass mill. Trained personnel inspect each lot to meet the most stringent U.S. dimensional and quality standards. PMC cases give smooth chambering and consistent reloadability.

In sum, we use maximum care in the production of all our components and you, the PMC customer, benefit from this care.

PMC takes exceptional care in the packaging and transport of all products. Solid wooden crates assure the safe arrival of your shipments. At the same time, our products are packaged in quantities which afford easy, economical distribution.

Stephen Szurlej

*These two examples of
close-up views
adequately demonstrate
the device; at the top is a
pore-by-pore look at a
salesman's smile, and at
right is a 30-30 rifle shell.
Both were shown many
times life-size.*

crowds in motion, Americans going nuts, middle-class life as it's lived—all strung up to the snapping point. I think he likes to see a crowd reacting in unison to some big wazoo. Years ago in the Universal commissary he would have Richard Nixon paged on the public-address system, just so he could watch a roomful of studio executives rise up from their tables and crane their necks.

For *1941*, he left his strong suit—the art of creating suspense—hanging in the closet; the movie races by too fast. Even with great character actors like Slim Pickens, Robert Stack, Warren Oates, Christopher Lee, Toshiro Mifune, Dan Aykroyd and John Belushi, the audience barely has a character to hold onto, or identify with. Bodies are continually being hurtled into unwieldy situations. An audience can be surprised only so many times. Chuck Jones, the animation genius behind Bugs Bunny and Road Runner, was consulted in the early design stages. Paying homage to cartoons and the old silent comedians in vivid color and pugnacious six-track sound transformed socko high jinks into something brutal. Spielberg learned about comedy on the most gargantuan canvas affordable.

The old comedic components — the slow burn, the double take, the dreadful anticipation, the witty rejoinder — were abandoned for the continuous slam-bang. For instance, if you knew before-hand that a paint factory was waiting behind closed doors, and that a careening army tank was heading straight for it, well, you'd be waiting. As it stands, the viewer is suddenly aware that a paint factory has materialized and this army tank is crashing through it. Or, in a cafe, a man's face pitching toward the counter suddenly meets a birthday cake. The mind registers these various happenings moments after the eye perceives them.

A few test previews of the film in Dallas and Denver caused some early nervousness (*Are they laughing?*), which led Spielberg to trim several minutes from the early reels. He might have cut too close to the bone. With the explanatory subplots cleaved away, we are left with a movie full of strangers. Later, he tried to find some consolation in John Williams' thunderous score. "Johnny's been overwriting over my overdirection over Zemeckis and Gale's overwritten script," Spielberg said.

Dan Aykroyd groped for lessons in this movie: "There were so many elements that had to be chopped out because the movie was so big, and he had to get it down to a manageable time. That's the big lesson, I think. Before you even start to walk onto a set or think about production, you've got to have a solid story that is so clear and vivid that there isn't much room to deviate and improvise, and therefore you don't have the selective process of cutting once it's all over. You have what you need in the can, and it enables the director to build on what was in the original script."

PHOTOGRAPH BY BONNIE SCHIFFMAN

Inescapable is the word for this image of the late John Belushi. The spread from Rolling Stone *magazine publicized his film* 1941.

★

H E CAME HOME FROM work one day last summer and parked the green Porsche. He entered his Beverly Hills aerie through the kitchen and stuck his nose into the icebox. Bertha, the respectable-looking housekeeper, said she'd make tea.

Spielberg's face was chiseled into a very serious structure. Gone was the youthful, cross-eyed goofiness. All of thirty-one years old, he looked a little more contemplative, a little more rugged. The lips were set. Decisions had left their mark. The jeans and the heavy dark blue shirt were built for comfort, but he'd lately replaced the striped jogging shoes with handsome brown boots.

It was a cavernous house in tasteful browns, high wooden beams above. It was something that might have been erected by a discreet, wealthy rancher. There was an antique billiard table, an outer-space sharpshooters game, a screening room under construction in the living room. He led the way to the den. It seemed to be the one cozy, overstuffed room in the mansion. Navaho rugs hung on the wall. A neat fan of Japanese architecture magazines lay on the table.

Dropping into a pillowy brown sofa, he chewed over a host of agonies. Maybe, he figured, it was time to get out of the dead spiritual air of Los Angeles. Get away from The Business. Maybe he should buy a ranch in Montana, like his business manager said. Standing on a Beverly Hills street corner, he couldn't hear Aaron Copland in his head. Slaving over a typewriter was much better in New York.

(When later I mentioned these hard lines of doubt to John Milius, he burst out laughing. "What it is, is he's getting *married*. He has to grow up and become a real *adult* now.")

The workday had been one long bout of the self-examination that a director gets in the sound-dubbing laboratories. To fine-tune the soundtrack, he had to watch the same ten-minute reel of film over and over again, and it's while doing just this job that the worry explodes and he sees what he did wrong and calls out the camera boys for another shot.

"I can't correct the overall conceptual disasters about *1941*," he said, putting his ankle on his knee, "but I can get little pieces here and there that I think will help speed the pace. If you can't do anything about it, then you're at the mercy of what comics call 'the death silence.' You expected a laugh and all there is is a hole."

This made me laugh. It was not the kind of boosterism a director usually serves up before a picture's release. His eyes lifted a little, ready for any understanding. He has, finally, an affable face. You find yourself agreeing with it.

I reminded him about his vow to laugh all the way to the final print.

"Yeah?" He smiled grimly. "Well, sur-

The face of the cheetah reflects an ancient and necessary relationship between predator and prey. The cheetah was born to kill, without pleasure and purely to satisfy its hunger. Its legendary speed (perhaps as high as 70 miles an hour for short distances) is a tool of its trade, for its prey is also fleet and keen. The cheetah must be fleeter and keener or it will not survive to bear young and perpetuate the species.

122 GEO

WHAT MAKES JACK RUN

What makes Jack Nicklaus tick? How did he become so competitive? What are the secrets of his success in life and business, as well as in golf? What are his goals? Contributing Editor Ken Bowden probed these subjects with Nicklaus in a lengthy tape-recorded session, and Nicklaus' candid observations offer a fascinating insight into his thinking on numerous topics. Bowden's questions and comments are in italics.

What does it take to be the best in the world at something?

Well, to be best at anything in the athletic field, obviously you must be pretty athletic. If you were going out to be No. 1 in business, I don't care how hard you work, if you're not smart you are not going to make it. In golf, if you're not a coordinated athlete, you are not going to get there.

When I was a kid, golf was never really a major sport. Today you find many good athletes being drawn to golf because of the great public awareness of the game. That's one reason I think you'll see more and more good golfers as time goes on. And better golfers, too—physically stronger, more finely coordinated athletes. Fellows like Hale Irwin—Hale Irwin is an all-round athlete. Gary Player is another good all-round athlete. I was a good all-round athlete. In fact, I think your very best modern golfers are nearly all fellows who would have been good athletes in any sport.

That's not all, though. There's not a really good golfer on the tour who is not mentally pretty sharp. You must be smart enough not only to understand how to play the game, but to understand *why* you play.

You were an all-round athlete in school and had a lot of physical equipment to play just about any sport. Why golf?

I played football in the fall, I played basketball in the winter, I played baseball in the spring, I ran track, I swam, I played handball, I played tennis—I played anything and everything.

But why golf in particular?

In football you have to have somebody to throw the football to and somebody to throw it back, right? I was a reasonably good quarterback, but the only thing I could do in football by myself was to learn how to

46   APRIL 1976/GOLF DIGEST

Closer-than-close looks at two aggressive specimens are given by these pages. The top spread from Geo *shows a hunting cheetah; the bottom pages show Jack Nicklaus, beard-stubbled and determined.*

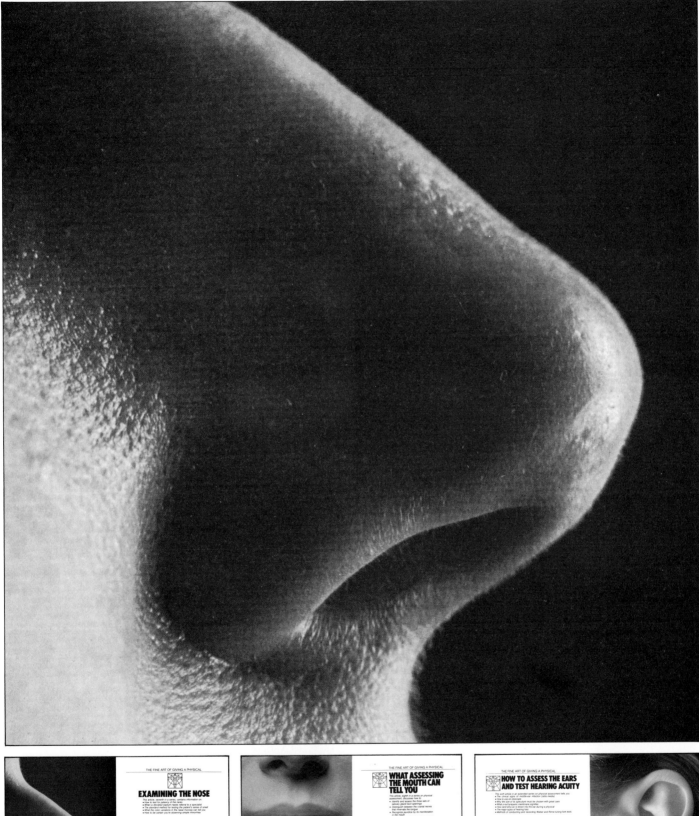

Stephen E. Munz

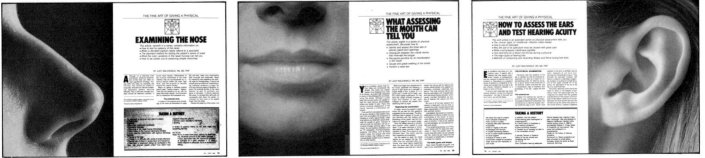

If close is shocking, even closer is more shocking. These spreads from RN *magazine dramatize three of a series of articles on giving a complete physical examination.*

THEY RECALL A 'BABY-FACED' MURDERER

Frank Walus, a 57-year-old Polish immigrant, lives on Chicago's Southwest Side, where he was once a Democratic precinct captain. He used to work on the assembly line at General Motors, collect coins and keep a pet pigeon in his garage. When Walus came to the U.S. in 1959, he told authorities that he had belonged to a wrestling club in Poland, but the court found he failed to report that he had also been a member of the Gestapo. For that omission Walus was brought before a judge two years ago and eventually stripped of his citizenship. At the hearing, 12 witnesses charged that Walus was a Gestapo agent —"the baby-faced Jew killer," said one man—who had murdered and brutalized Jews in the ghettos of Czestochowa and Kielce. One witness, David Gelbhauer, an Israeli locksmith, said he saw Walus shoot a woman, two little girls and a Polish partisan in cold blood. He also described dragging battered people from an interrogation room where Walus allegedly had beaten them. Walus, who wore a tiny American flag on his lapel at the hearing, denied the crimes and is now appealing the decision. Since the proceedings began, he has been attacked with Mace and threatened. To meet continuing legal costs, he has placed three mortgages on his house. Walus sees himself as the victim of a conspiracy. In addition he protests that only Himmler can be held accountable for Nazi crimes. "There are always war crimes," Walus says in broken English. "From lieutenant on down, he was following order[s], yelling, 'My God, my God, if I wouldn't do that, they'd kill me.'"

Frank Walus is alleged to have killed many Jews in the Polish ghetto of Czestochowa.

Witness David Gelbhauer (above) testified that Walus (right) was a Gestapo agent.

This spread from a Life story about fugitive ex-Nazis in America uses abrupt change of scale to bring us face-to-face with an accused mass murderer.

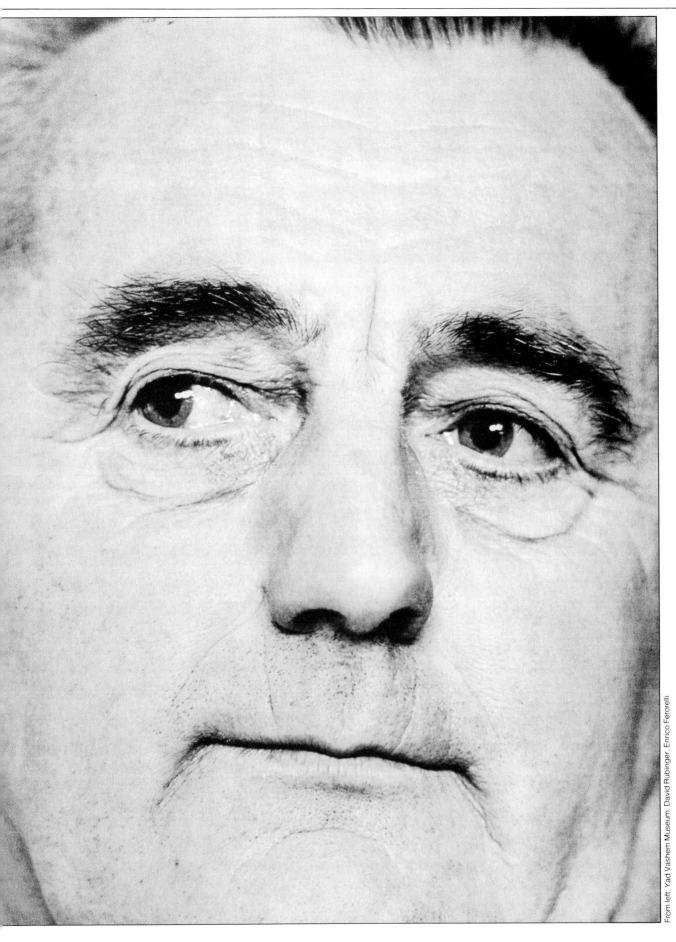

SPECIAL BIG MONEY SECTION   In football, quarterbacks are ranked by touchdown passes. Baseball's sluggers are measured in home runs. Hockey goal-tending supremacy is decided by the goals-against column. But when it comes to pro golf, the eye-grabbing statistic is money . . . big money. Most other figures are lost in a wasteland of tiny type. The big news is that Jack Nicklaus won $320,542 in a single year. That Bobby Nichols bagged $50,000 by winning the Westchester Classic in overtime. That Big Four champions Nicklaus, Weiskopf, Miller and Aaron played 36 holes for $50,000 in the World Series of Golf. Money, money, money . . . the name of the game.

Although sitting atop the highest greenback stack of all, Nicklaus prefers recognition for his record 14 major championships rather than his millionaire status. He reminds that the Hogans, Hagens and Nelsons are swamped under when money is the measuring stick. After all, superkid Lanny Wadkins has banked more in two years than Hogan did in his career. Matter of fact, Weiskopf won almost as much during a nine-week summertime hot streak. But no matter how constant his plea to play down the earnings total, Nicklaus is stuck with being rated by the money he wins. America's capitalistic society identifies with money. The golfing public admires the men who make fortunes playing a game. On the following pages you get a look at the evolving influence of money—for good and bad—on the tour. There's the story of a $500,000 tournament, the richest in history, that cannot attract all the game's superstars even with a $100,000 top prize. There's also the tale of Johnny Miller, the handsome youngster who was prepackaged and merchandised successfully by an enterprising business manager before he even established himself on the tour. And there's golf's expanding "Millionaires Club" in a surprise, but appropriate, setting.

NOVEMBER 1973/GOLF DIGEST   49

Large type

Type is not pictorial, strictly speaking, but it can be a strong graphic device when used for a larger than normal headline. It can accent and emphasize important parts of the material but, like any good seasoning, big type should be used sparingly.

This page-size dollar sign announced a special report section on golfers who had passed the million dollar earning mark. The symbol was repeated in various ways throughout the fourteen pages of the section.

In the example at right, large and colorful type was the unifying and dramatizing element in the header page for a story about Sam Snead's sixtieth birthday.

Enlarged type changed the dynamics of this headline to help convey the author's surprise.

If Hollywood were to script the life and times of Ralph Guldahl, they'd probably have him making a pact with the Devil. In return for four glorious years of golf, Guldahl would agree to forfeit the remainder of his career.

And the Hollywood version would be almost as credible as the real thing. Ralph Guldahl is the Amelia Earhart of golf. For a brief time he flamed like a shooting star. Then he vanished from the public gaze. In a span of three years he won the U.S. Open twice, the Western Open three times, the Masters once, and twice was runner-up in the Masters. Then in one year he went from *Who's Who* to *Who's He?* It was as though Babe Ruth had suddenly forgotten how to hit, or Gordie Howe to shoot a hockey puck, or Bill Russell to block a basketball shot.

In modern times only two other golfers have scored back-to-back triumphs in the U.S. Open. Two of them, Bobby Jones and Ben Hogan, are major historical figures of the game. Ralph Guldahl, the Open champion in 1937-38, left only his spike marks on the pages of history. But those who witnessed Guldahl in his prime still regard him with awe. Guldahl, as a past Masters champion, was making his perennial appearance on the fairways at Augusta National last spring *(photo on opposite page)* when he caught the attention of Fred Corcoran.

"He was the greatest golfer in the world," mused Corcoran, the first tournament director of the PGA, "and then lost it overnight. He woke up one morning and it was gone."

According to Guldahl, that's only a part of the truth. Guldahl knows that the passing years have enveloped him in an aura of mystery. He is aware that people speculate on why he quit tournament golf at the top of his game, but he has never cleared up the confusion. He has been almost a recluse since his retirement.

"I didn't lose my swing," he protests. "I lost my incentive. If I had continued to play, maybe I would have been unable to take the clubhead back. But I quit. I guess I am inherently lazy; I figured it was time to settle down."

He settled in California with his wife, Maydelle, and his son, Ralph, Jr. The elder Guldahl had been a fledgling of 19 when a pretty girl named Maydelle Laverne Fields came to him in Dallas for golf lessons. Soon the teacher was displaying more than a casual interest in his pupil. "I took money for a few lessons," he recalls, "then I didn't take any more."

They were married in 1931 and Ralph, Jr., arrived in 1935. The youngster was a powerful influence on Guldahl's decision to abandon tournament golf. "It wasn't much of a life for the kid," says Ralph, Sr. "He was in and out of military schools while we were traveling."

Now the father, at 59, is director of golf at Braemar Country Club in Tarzana, a bedroom community of Los Angeles. The son, once a golf pro, is associate editor of Hot Rod Magazine.

As you might suspect, Tarzana was the home of the late Edgar Rice Burroughs, who wrote about a jungle swinger and made work for Johnny Weissmuller, Buster Crabbe and other well-muscled actors. Guldahl commutes daily from his attractive home in Sherman Oaks to Tarzana, a few freeway miles away in San Fernando Valley.

At times he dreams of past glory. He can recall every shot of his victories in those Opens at Oakland Hills and Cherry Hills, and he laughs if you kid him about his habit of combing his hair after hitting a shot. Golfers a bit long in the tooth can picture Guldahl standing in the 18th fairway at Oakland Hills the last day, running a comb through his dark hair before he came to the green to accept the cheers of the crowd. "I knew I had won the Open," he says, "and I figured they'd be taking pictures of me. I wanted to look nice for the cameras. I had a good head of hair in those days and I was proud of it. My wife always said she married me because I had long, curly hair."

One who watched impatiently while Guldahl combed his hair was Sam Snead. Snead had finished ahead of Guldahl and seemed a sure winner. Then Guldahl defeated him by two strokes with a record score of 281. That was the beginning of Snead's frustration in the Open, a trauma that has lasted throughout his career.

"Sam always figured he didn't win four or five Opens because I came from behind and beat him out that year," says Guldahl. Not that Guldahl rejoiced over Snead's mis-

HE
WON
THE
OPEN?

TWICE?

Ralph Guldahl came out of the woodwork to win amazing
back-to-back victories in the U.S. Open, then
mysteriously returned to obscurity. Here's why.

BY JACK MURPHY

52   DECEMBER 1971/GOLF DIGEST

Leonard Kamsler

SNEAD AT 60

The man whose face appears on this page buys his straw hats (two dozen a year) from a supplier in Mallory, N.Y., eats creamed chipped beef on toast for breakfast, owns and is adored by a huge black Doberman named Adam— and needs about as much introduction as Richard Nixon. He is, of course, Sam Snead, the reigning Club Professional champion of the U.S., a man whose golf swing has been coveted by millions of golfers covering several generations. This year, three weeks past his 60th birthday, as the PGA Senior champion, he met the British Senior titlist for the World Senior Professional crown. He still competes regularly and well on the PGA tour. Snead's remarkable talent and longevity are dealt with in a special section that covers the next 16 pages, an uncommon tribute to an uncommon performer. The *Golf Digest* professional panel of experts analyzes in detail why Snead's swing, captured in recent color sequence photos, has stood up to the demands of top professional competition for so long. Then, Cal Brown offers a revealing personal study of the man behind the golfer, explaining why Snead has not stopped competing and may never stop. Finally, Snead, a long-time member of our staff, describes for the first time the key concepts he uses to check his swing when it gets off track. Like his swing, Snead's reference points are effective and simple, and in his simplicity there is beauty enough for any golfer to treasure.

115

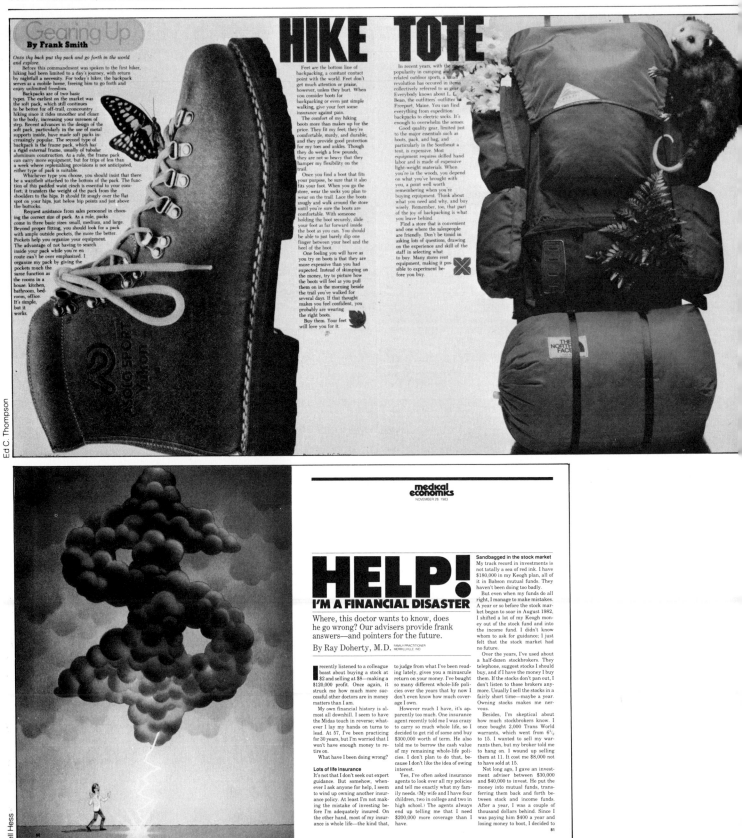

Gearing Up

HIKE TOTE

By Frank Smith

Onto thy back put thy pack and go forth in the world and explore.

Before this commandment was spoken to the first hiker, hiking had been limited to a day's journey, with return by nightfall a necessity. For today's hiker, the backpack serves as a mobile home, freeing him to go forth and enjoy unlimited freedom.

Backpacks are of two basic types. The earliest on the market was the soft pack, which still continues to be better for off-trail, crosscountry hiking since it rides smoother and closer to the body, increasing your sureness of step. Recent advances in the design of the soft pack, particularly in the use of metal supports inside, have made soft packs increasingly popular. The second type of backpack is the frame pack, which has a rigid external frame, usually of tubular aluminum construction. As a rule, the frame pack can carry more equipment, but for trips of less than a week where replenishing provisions is not anticipated, either type of pack is suitable.

Whichever type you choose, you should insist that there be a waistbelt attached to the bottom of the pack. The function of this padded waist cinch is essential to your comfort; it transfers the weight of the pack from the shoulders to the hips. It should fit snugly over the flat spot on your hips, just below hip points and just above the buttocks.

Request assistance from sales personnel in choosing the correct size of pack. As a rule, packs come in three basic sizes: small, medium, and large. Beyond proper fitting, you should look for a pack with ample outside pockets, the more the better. Pockets help you organize your equipment. The advantage of not having to search inside your pack isn't over emphasized. I organize my pack by giving the pockets much the same function as the rooms in a house: kitchen, bathroom, bedroom, office. It's simple, but it works.

Feet are the bottom line of backpacking, a constant contact point with the world. Feet don't get much attention or praise, however, unless they hurt. When you consider boots for backpacking or even just simple walking, give your feet some insurance against pain.

The comfort of my hiking boots more than makes up for the price. They fit my feet; they're comfortable, sturdy, and durable, and they provide good protection for my toes and ankles. Though they do weigh a few pounds, they are not so heavy that they hamper my flexibility on the trail.

Once you find a boot that fits your purpose, be sure that it also fits your foot. When you go the store, wear the socks you plan to wear on the trail. Lace the boots snugly and walk around the store until you're sure the boots are comfortable. With someone holding the boot securely, slide your foot as far forward inside the boot as you can. You should be able to just barely slip one finger between your heel and the heel of the boot.

One feeling you will have as you try on boots is that they are more expensive than you had expected. Instead of skimping on the money, try to picture how the boots will feel as you pull them on in the morning beside the trail you've walked for several days. If that thought makes you feel confident, you probably are wearing the right boots.

Buy them. Your feet will love you for it.

In recent years, with the rising popularity in camping and related outdoor sports, a small revolution has occured in items collectively referred to as gear. Everybody knows about L. L. Bean, the outfitters' outfitter in Freeport, Maine. You can find everything from expedition backpacks to electric socks. It's enough to overwhelm the senses.

Good quality gear, limited just to the major essentials such as boots, pack, and bag, and particularly in the Southeast a tent, is expensive. Most equipment requires skilled hand labor and is made of expensive light-weight materials. When you're in the woods, you depend on what you've brought with you, a point well worth remembering when you're buying equipment. Think about what you need and why, and buy wisely. Remember, too, that part of the joy of backpacking is what you leave behind.

Find a store that is convenient and one where the salespeople are friendly. Don't be timid in asking lots of questions, drawing on the experience and skill of the staff in selecting what to buy. Many stores rent equipment, making it possible to experiment before you buy.

Ed C. Thompson

---

medical economics
NOVEMBER 28, 1983

HELP!
I'M A FINANCIAL DISASTER

Where, this doctor wants to know, does he go wrong? Our advisers provide frank answers—and pointers for the future.

By Ray Doherty, M.D. FAMILY PRACTITIONER MERRILLVILLE, IND

I recently listened to a colleague boast about buying a stock at $2 and selling at $8—making a $120,000 profit. Once again, it struck me how much more successful other doctors are in money matters than I am.

My own financial history is almost all downhill. I seem to have the Midas touch in reverse; whatever I lay my hands on turns to lead. At 57, I've been practicing for 30 years, but I'm worried that I won't have enough money to retire on.

What have I been doing wrong?

Lots of life insurance

It's not that I don't seek out expert guidance. But somehow, whenever I ask anyone for help, I seem to wind up owning another insurance policy. At least I'm not making the mistake of investing before I'm adequately insured. On the other hand, most of my insurance is whole life—the kind that,

to judge from what I've been reading lately, gives you a minuscule return on your money. I've bought so many different whole-life policies over the years that by now I don't even know how much coverage I own.

However much I have, it's apparently too much. One insurance agent recently told me I was crazy to carry so much whole life, so I decided to get rid of some and buy $300,000 worth of term. He also told me to borrow the cash value of my remaining whole-life policies. I don't plan to do that, because I don't like the idea of owing interest.

Yes, I've often asked insurance agents to look over all my policies and tell me exactly what my family needs. (My wife and I have four children, two in college and two in high school.) The agents always end up telling me that I need $200,000 more coverage than I have.

Sandbagged in the stock market

My track record in investments is not totally a sea of red ink. I have $180,000 in my Keogh plan, all of it in Babson mutual funds. They haven't been doing too badly.

But even when my funds do all right, I manage to make mistakes. A year or so before the stock market began to soar in August 1982, I shifted a lot of my Keogh money out of the stock fund and into the income fund. I didn't know whom to ask for guidance; I just felt that the stock market had no future.

Over the years, I've used about a half-dozen stockbrokers. They telephone, suggest stocks I should buy, and if I have the money I buy them. If the stocks don't pan out, I don't listen to those brokers anymore. Usually I sell the stocks in a fairly short time—maybe a year. Owning stocks makes me nervous.

Besides, I'm skeptical about how much stockbrokers know. I once bought 2,000 Trans World warrants, which went from 6½ to 15. I wanted to sell my warrants then, but my broker told me to hang on. I wound up selling them at 11. It cost me $8,000 not to have sold at 15.

Not long ago, I gave an investment adviser between $30,000 and $40,000 to invest. He put the money into mutual funds, transferring them back and forth between stock and income funds. After a year, I was a couple of thousand dollars behind. Since I was paying him $400 a year and losing money to boot, I decided to

Lowell Hess

The accented word "Help" creates a strong hook for the reader and puts more emotion into the headline.

SLEEP

A pound of feathers and a pound of lead may weigh the same, but the feathers will make a better sleeping bag and keep you warmer.

Actually, high-quality sleeping bags are made not of feathers but of down from geese or ducks. Synthetic fiberfills that are lightweight, like PolarGuard, also make excellent bags. The virtue of both down and PolarGuard is in their ability to loft, or fluff up, thereby creating a space of dead air that provides a barrier between the heat of your body and the cold night air. With a properly designed and stitched bag, three to four pounds of down will keep you warm when the temperature is zero.

In considering the purchase of a sleeping bag, think about the conditions in which it will be used. Sleeping bags are usually rated by the coldness of expected camping conditions; if you see no rating, a knowledgeable salesperson can be of help. Down is more efficient than a synthetic fill, but it is also more expensive. PolarGuard is bulkier, which may concern you if your pack has a limited compartment for your sleeping bag, but it also has the advantage of insulating even when wet, which down does not do.

Generally, bags come in one of three shapes: rectangular, a modified rectangular cut, and the mummy, which is the most efficient in keeping you warm. The way in which the bag has been sewn is important: stitching should be clean and close together, since any loose stitches where the inner shell is attached to the outer shell will cause a heat loss while you sleep. Zippers should work easily from inside or outside the bag, and you should be able to zip from either end, allowing you to create ventilation at the foot of the bag on warm nights.

As with most quality gear, sleeping bags worth the money are expensive. On that really cold night in the woods, however, you will appreciate every dollar you spent.

WITH IMAGINATION, NERVE AND ENOUGH MONEY, ANYBODY CAN PUT A ROCKET INTO SPACE. OR TRY.

SPACE INC.

Private enterprise began its space odyssey two years ago with a fiery false start. The setting was a Texas pasture that served as a launching pad for Percheron, a 55-foot rocket named for the reliable French workhorse. It was the forerunner of a new breed of low-cost rocketry fueled by the profit motive. But it exploded on the pad (photo at right), scattering space-age debris amid scores of flaming cow pies.

The experience has not deterred a handful of entrepreneurs who are determined to make a profit by delivering people and cargo to space. Some are building expendable boosters with parts and expertise supplied by the aerospace industry. Others are committed to reusable rockets based on promising but unproved technology. The race to become the Federal Express of the space-freight business has begun in earnest.

New communications satellites destined for geosynchronous orbit will be the most profitable cargo. Other potential markets include earth-sensing satellites that supply data used by crop forecasters and oil prospectors; orbital factories for processes enhanced by extremely low gravity; and space-based disposal of nuclear wastes.

These markets should be the bread and butter of NASA's space-shuttle fleet. But the shuttle has a bookkeeping problem. NASA now charges customers a total of $42 million for a single mission that costs taxpayers more than $200 million. The fare per mission will increase to $71 million in 1985. Nobody knows what it will be in late 1988, when the subsidy ends and the shuttle must cover operating costs. But it's likely to be astronomical.

Today's space entrepreneurs, unburdened by NASA's bureaucratic waste, overhead and top-shelf standards, are quoting prices that are pocket change compared with the space agency's subsidized rates. Can they deliver? The odds are certainly against them. But the few who succeed should see the world beat a path to their launching pads.

Article by Junius Ellis • Photos by Norman Seeff

46 GEO

Focus on the reader

Just as it is more convincing for a person to look directly in your eyes when he or she speaks, it is more attention-producing when the subject of a photograph or drawing talks, looks, or gestures toward the reader. The cover of the annual Bum Steer issue of Texas Monthly, *below, gives the reader a beefy Bronx cheer.*

JANUARY 1984

$2.00

Candy Montgomery: The Adulteress Who Beat a Murder Rap
Throw Those Glasses Away: Fixing Your Eyes by Surgery

TexasMonthly

1984 BUM STEER AWARDS!

A Great Big Raspberry to Miss Texas, the Aggies, "Tootsie" Whitmire, Carolyn Farb, Mean Joe Greene, Cullen Davis, and the Dallas Cowboys Cheerleaders.

Tom Curry © 1984 *Texas Monthly*

Ray Woolfe

A Messerschmidt 109 fighter from the Confederate Air Force, a vintage plane collector's club in Texas, heads directly out of the page.

The physical conditioning of a hockey goalie is a year-round concern, so Hockey magazine photographed a player in full regalia in a summer meadow. This picture has other things going for it, but is more arresting because the goalie is looking straight at the reader.

Kipling might have said it this way. Things that you learn on the grass in the sun will help you a heap on the ice in the cold. Goaltending begins with confidence and ends in quickness and toughness on the ice. Working and planning in the off-season gives a serious, competitive goalie a definitive advantage. This means more than having some winger firing pucks at you.

Let's analyze what can be done to strengthen confidence and increase quickness, even when the ice has trickled down the drain.

It is often suggested that goalies do not need to be as firmly conditioned as other hockey players. Don't you believe it. Not only is the goalie on the ice for the whole game—the demands on his alertness are constant, and concentrated action can be frequent. When a goalie is in poor condition, mental fatigue takes over. Reactions slow down. The goalie looks like he is playing under water. One slip, one goal. Two errors, two goals. A goalie's mistakes appear on the scoreboard.

A seasoned player creates his confidence with a sound and rugged conditioning program. Jogging and bike riding are solid general activities for summer conditioning. Get interesting variety into your training—jog backwards, actually sprint backwards often. When the season is six weeks away, start charging up hills, leaping up

Dave Kilgour is the head hockey coach and jayvee lacrosse coach at Wilton High School, Wilton, Conn. In the summer he is an instructor at Paul Lufkin's Darien Hockey School. He played goalie at Bowdoin College.

stairs, and running up bleachers.

Play ping-pong every chance you get—play with another goalie—put a paddle in each hand, or better still, play with your hockey gloves on. Your arms can be strengthened by adding weights around the wrists, even add weight to your ankles. Then play wild, fast ping-pong or handball . . . and wear your mask while you play.

One way to find weights is to check a gasoline station or tire store. Usually they have worn tire balancing weights that might otherwise be thrown away. These can be sewn into a band made from an old shirt or sheet, then taped on to your wrists or forearms. Again, be sure to put weight on your ankles, too.

Some goalies switch to the lacrosse net and others play behind the plate in baseball during the spring and summer months. Good—but the timing is different, and the ball originates in a different way in these sports. However both will increase your quickness and you do become used to donning the extra padding that all backstops must call their own.

A major drawback with most young goalies is keeping the puck clearly in focus at all times. There's a way to adjust that. A simple series of eye exercises can strengthen the eye muscles and cut down the time it takes to focus on the puck. Try sitting in a chair then roll your eyes, checking to see how far above, below, left, and right you can see without moving your head. This is done best alone, for five minutes a day. Later, alternate focusing on a close object, then on a far away object.

Out on the grass (for this drill get a friend to help you), throw a

ball to your partner, do a somersault and catch the return. Super quick goalies can do this with a tennis ball against a wall. But you have to be swift.

Here's another way goalies can work to improve over the summer. Do the split. Try it this way. Stand with your legs spread as far apart as possible. Now slowly try to lift your heels, bend at the knee, or roll your hips. Get up without lifting your feet, or jumping.

Finally, a goalie should become a thinking man, working over his technical bag of tricks—always plotting for next season. Slowly figure out each of your regular moves . . . talk them out, practice them. Can you stack your pads at both sides? Half split, full split both directions? Butterfly? Where does your stick go during each move? How well do you play the angles? What's your record on breakaways? Anticipate everything and you'll be prepared for anything. How many rebounds can you handle before they do you in? How quickly can you put the puck into either corner?

Goalkeeping is a lonely position . . . it takes special talents, and a certain kind of mind and body. A goalie must be a person who is absolutely determined to keep the puck out of the net. A part of it all is physical—good conditioning, reflexes; another part is experience, skill, and courage; knowing what to do and how to do it. But the significant part of goalkeeping is confidence in yourself and your ability. You gain this confidence by preparation. In the winter your summertime scheming and conditioning will all come to bear. Defensive, maybe, but goalies win hockey games. Ask anyone who has seen a team crowd around after a game.●

Goalie training is a year-round job

By DAVID KILGOUR

John Newcomb

ION TIRIAC: THE TENNIS BANDIT

He's surly, tough and, with that scowl, perfectly cast as one of the bad guys of the sport. But is that bad guy image just a facade? Not really.

By RICHARD R. SZATHMARY

Coming on court at Nassau Coliseum for a World Team Tennis (WTT) match against the New York Sets, the visiting Boston Lobsters occasion little fan comment—even in a league environment that rather shamelessly and constantly seeks to whip up a little excitement from the stands. Kerry Melville Reid is perky/cute/sharp, Valerie Ziegenfuss outright luscious, Wendy Turnbull reliable and even, Raz Reid surfer-handsome, Bob Hewitt imposingly bald, and that's about as far as they go in terms of visual interest.

Even the guard holding the door for

66 TENNIS/September 1975

John Newcomb

"Count Dracula," Ian Tiriac of Rumania, is known for his irritability and violent temper, which are effectively conveyed by this article opener.

WHY THE OPEC BUBBLE BURST

An analyst who foresaw the recent price drop tells where so many "experts" went wrong.

By Arlon R. Tussing

At the end of 1981, oil prices entered their greatest decline in half a century, but many industry executives and forecasters cautioned that the attendant glut was a temporary phenomenon. Despite many signals to the contrary, that opinion persisted through 1982, with Occidental's Chairman Armand Hammer and others predicting $100-per-barrel oil in 10 years. The scarcity mentality that fed the price leaps of 1974 and 1979, plus a belief that greedy Middle Easterners controlled the world oil market through OPEC, gave such assumptions nearly axiomatic status during the past decade. But the exporting nations' boast that "oil in the ground is a better investment than money in the bank" is turning out to be so much wishful thinking.

The truth is that neither an end to the recession, nor OPEC attempts at production quotas, nor continued wars in the Middle East, will long be able to shore up a sagging crude-oil market. It is, indeed, because oil prices climbed so rapidly and so high in the 1970s that they are now almost certain to keep falling—perhaps as steeply and as far as they rose. The forces that led to the enormous petroleum price increases of the past decade

This author was chief economist for the U.S. Senate Committee on Energy and Natural Resources from 1972 to 1976. He is now professor of economics at the University of Alaska and president of ARTA Inc., accounting firm in Seattle. This article is condensed, with permission, from the author from the Public Interest, Number 70, Winter 1983. Copyright © 1983 by National Affairs Inc.

160 MEDICAL ECONOMICS/MAY 2, 1983

Stephen E. Munz

The despairing OPEC sheik stares forlornly at the reader as he puts the gas pump to his temple.

The gaze of anyone or anything can be riveting —even this gigantic African bullfrog.

tennis

$1.00
MARCH
1980
35392

LARGEST PAID
CIRCULATION
OF ANY
TENNIS
PUBLICATION

3 volleys to make you practically invincible at the net

So you think it'd be fun to house a tournament player

Peter Fleming: the Paul Bunyan of pro tennis

VIC BRADEN: LET ME MAKE YOU FAMOUS BY FRIDAY!

How to hit an approach shot, by Arthur Ashe

Key rules you may not know about – but should

Dream vacation: what $800-a-week will get you in tennis luxury

Are age rules unfair to some juniors?

What the '80 Masters proved

Mark Kozlowski

Vic Braden, the clown prince of tennis instruction, buttonholes the reader with his promise of instant improvement.

In deliberate emulation of the famous Uncle Sam of James Montgomery Flagg's World War I recruitment poster, a laboratory supervisor uses the same gesture to warn against waste.

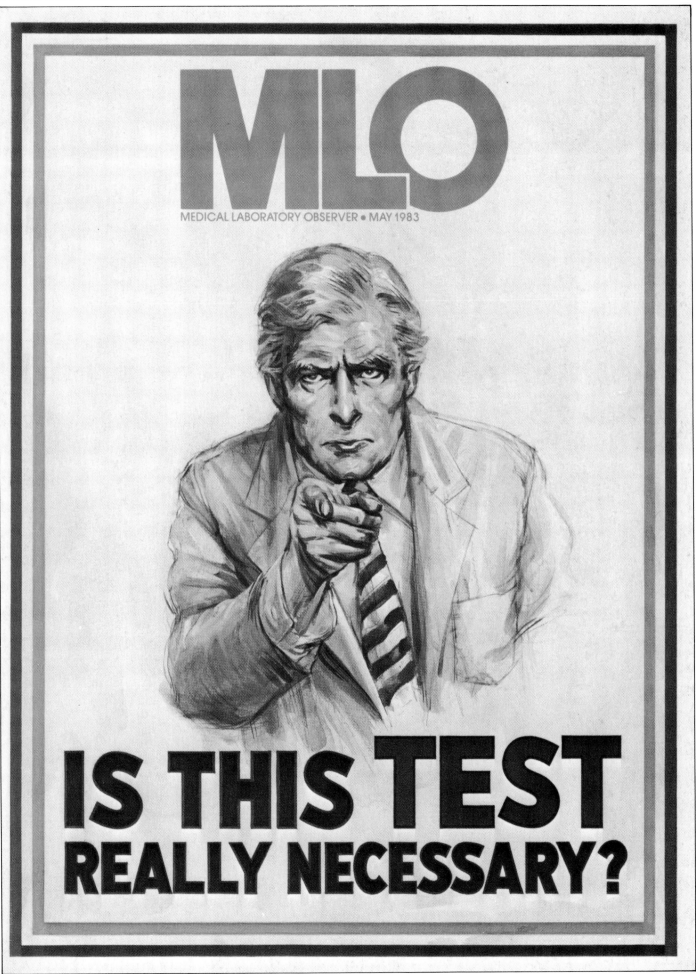

Unusual point of view

An eye-catching act can be made with a picture that looks at the subject from an unfamiliar angle or that seems to have been taken from an impossible position, a vantage point with no room for either camera or operator.

This shot makes it seem that the camera was operated from the bottom of a golf hole. Actually, the shot is a clever mock-up created with black paper taped to the photographer's window.

The all-powerful and threatening judge is made to seem even more menacing by the low camera angle, which makes the bench seem like an unreachable mountaintop.

Special Putting Section

Let's suppose there is a Golf Fairy. Let's pretend further that this lovely sprite visited your castle. "I shall grant you one wish," says the Fairy Golfmother. "I shall make you the greatest performer on earth with one golf club of your choosing. Which club do you select?" Don't hesitate, friend. Grab your putter and toss it to this magical mom. There is no other club in your bag as important as your putter. If you ever play a par-72 course in "perfect regulation," the putter will hit half your shots. Putting . . . . it's the quickest way to drive even the finest of golfers crazy. It has destroyed tour players who otherwise could match $200,000-a-year performers shot for shot. Likewise it has saved the careers of others. To lead off this special 16-page putting package, the touring professionals themselves rate the finest putters on the circuit. Next, you'll get some "how to" putting tips from Jack Nicklaus, a man who just might have the Golf Fairy locked in his closet. There's also a guide to purchasing a putter—this most important golfing tool designed in countless shapes and styles. If you're looking for a putting legend, you can read about George Low. He never won a Masters or a U.S. Open, but he has made a small fortune on the practice green and has given lessons to some of golfing's greatest names. The putter purchasing guide and Low's putting instruction are excerpted from our new book, "All About Putting." Oh yeah, one more tip . . . . don't sit around too long waiting for the Golf Fairy to show up .

Ralph Breswitz

---

medical economics
SEPTEMBER 13, 1982

FLUNK THIS QUIZ, AND YOU COULD END UP IN COURT

One of the best hedges against malpractice suits is knowing where the risks lie. Do you?

By James Griffith, J.D.

The average malpractice jury award against doctors has increased more than two and a half times since 1975, from $94,950 to $244,610. In cases where doctors and hospitals are co-defendants, awards have almost tripled; they average $450,130. One doctor in four will be a malpractice defendant at some point in his lifetime.

These numbers are scary enough, but consider this: More and more often, I find myself representing physicians who have made no medical error. Although practicing impeccable medicine is still the best defense against becoming a malpractice defendant, it isn't enough. Today you need to be able to recognize a malpractice risk before you take it.

I've devised this quiz, which consists of 10 multiple-choice questions, as a sort of malpractice refresher course. Your score should give you some idea of how well-informed you are.

If you get eight or more correct, you can consider yourself medico-legally superior. If you score between five and seven, you have a fair knowledge of the subject. If you score less than five, you'd better do some boning up.

1. You've completed a Caesarean delivery and now you're about to perform a tubal ligation. You observe that both ovaries are diseased, and since you know they will eventually have to come out, you perform a bilateral oophorec-

THE AUTHOR, a Philadelphia attorney who specializes in defending doctors and hospitals, is an editorial consultant to MEDICAL ECONOMICS.

76

Stephen E. Munz

GOLF DIGEST

35710 $1.25
October 1979
Largest
circulation
of any
golfing
publication

How to swing in balance, by Tom Watson

Why golf's TV ratings are sagging this year

Arnie turns 50!

Is Jim Rice of the Red Sox the longest hitter in golf?

Tom Watson

John Newcomb

Here it looks as though the ground had suddenly fallen away from top golfer Tom Watson as he was teeing up his ball. Actually, he was carefully balanced on an inch-thick sheet of Plexiglas supported by sawhorses.

Beware of the obvious NEAR-D VICTIMS

Make no assumptions: A 'lifeles patient may not be dead; a 'rec may be in danger . . .

BY ANITA DONAHUE, RN

When it comes to drowning, things aren't always what they seem. A near-drowning victim who appears fully recovered may die of complications two days later; one who looks dead when he's first pulled from the water may still be revived by aggressive, unremitting CPR. The familiar clues that usually help in assessing physical condition—skin temperature, pulse, dilation of pupils, cyanosis—can be downright *misleading* when near-drowning is involved.

THE AUTHOR is head nurse in the Emergency Department of South Miami (Fla.) Hospital.

Produced by Marcia Ringel Barman

Shig Ikeda

Neither subject nor camera should be where this picture was taken—under the surface of a swimming pool. The camera was inside a lucite box and the baby was dropped in front of the lens. (The model was the calmest member of the shooting party; she enjoyed being dunked.)

...WNING

tim

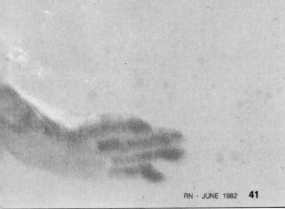

...ssible to say for sure ...ent has succumbed until ...n to a hospital and treat- ...now "obvious" his condi- ...ar.

...the scene of a drown- ..., Emergency Medical ...MTs) in the ambulance, ...d other personnel in the ...ust all proceed on the ...at prompt, correct, and ...suscitation will make a ...despread knowledge of ...ct alone would probably ... any more survivors. Ac- ...e National Safety Coun- ...ownings took place in ...or some reason, that ...ns about the same from

...ly, far too many deaths ...ing are declared soon

after recovery from the water. As you'd expect, duration of submersion is often the deciding factor. But even that can fool you: Many recoveries have been reported after submersion for half an hour or longer. The cardinal rule is CPR first, continued until the victim responds or a physician declares him beyond hope of recovery.

Even after successful CPR, the victim may still be in danger. Resuscitation followed by death from shock lung, cardiac arrest, or chemical pneumonitis caused by the aspiration of gastric contents is common. A number of names, including *secondary drowning, late drowning, delayed drowning, post-immersion syndrome, late pulmonary edema,* and *parking-lot drowning,* describe post-resuscitation problems. A typical scenario: Rescuers bring the victim out of the water, restore him to full consciousness, and then abandon treatment. He dies at home, or in the

RN · JUNE 1982 **41**

Emphasis on telling detail

Often the most dramatic way of telling a pictorial story is to show not sweeping panoramas but, rather, a close look at a comparatively small part of the action, a narrow but significant slice of the main scene. The cover below shows the skillful use of such a detail; the eye of the aging surgeon tells the story in a way that lingers in the mind.

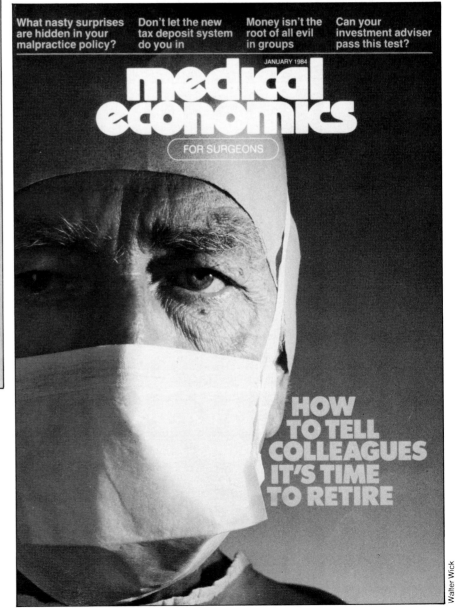

What nasty surprises are hidden in your malpractice policy?

Don't let the new tax deposit system do you in

Money isn't the root of all evil in groups

Can your investment adviser pass this test?

JANUARY 1984

medical economics

FOR SURGEONS

HOW TO TELL COLLEAGUES IT'S TIME TO RETIRE

Walter Wick

127

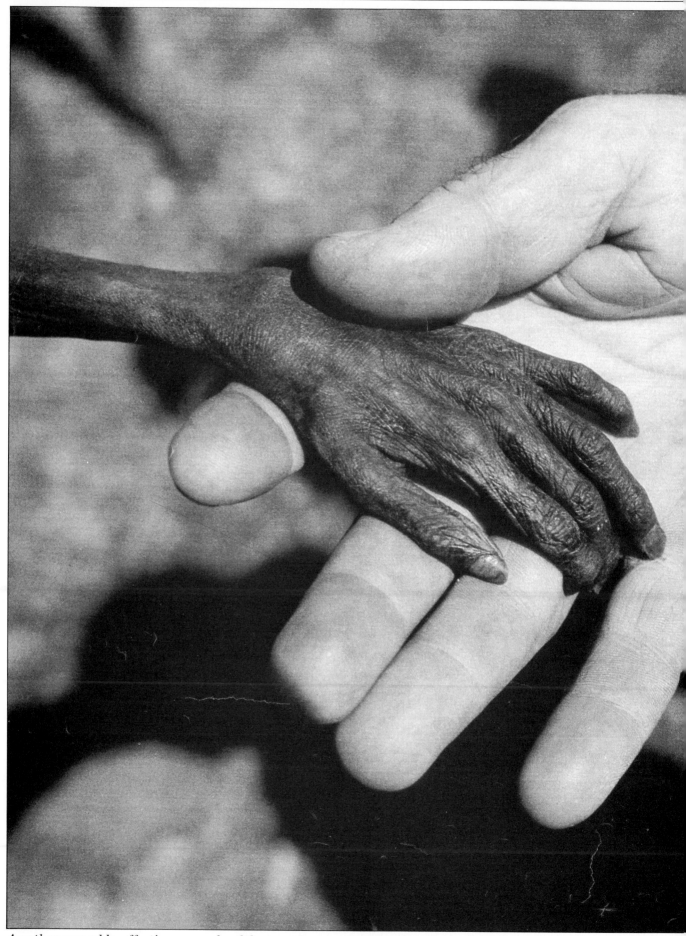

Another superbly effective example of the use of a small detail in a large picture is this spread from a Life *magazine story about a famine in Uganda. The sm[a]ll hand belongs to a starving child.*

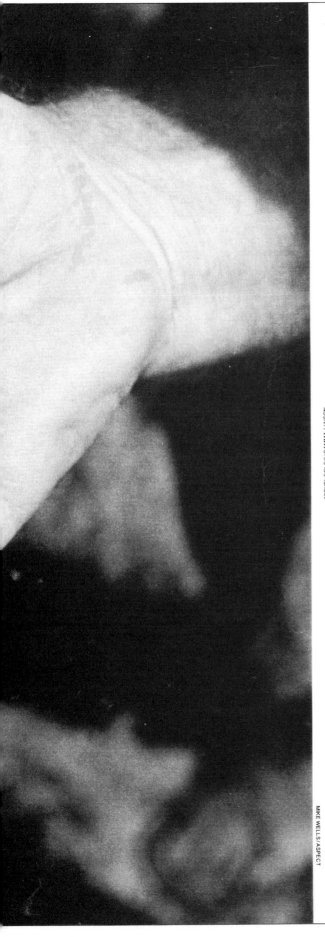

MIKE WELLS/ASPECT

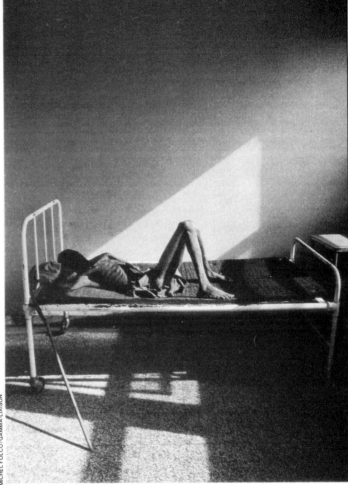

MICHEL FOLCO/GAMMA-LIAISON

Drought's Harvest in Uganda

The scenes are heartrendingly poignant, but do not be misled—it is mere succor that is being offered here, not sure rescue. The white priest's hand is indeed extended in compassion, yet only a miracle could save the famished child, now frail and wasted by the ravages of kwashiorkor, chronic malnutrition. Likewise, the hospital bed possesses no magic for the patient collapsed after months of starvation.

This is Uganda, nightmare land where drought has plagued 13 million people still adrift in the chaos left by the mad tyrant Idi Amin, who fled last year. Uganda is sui generis in a continent where the macabre remains common enough. Life in a place once described by Winston Churchill as "the pearl of Africa" is often just a kind of semideath. In the Karamoja region, where the drought is most severe, more than 20,000 people have starved this year alone. Here the Karamojong tribesmen, brigands who have made stealing a way of life for centuries, have traded spears for tommy guns and use them to raid vehicles bringing relief supplies. When the government sends in militia, the soldiers often join the pillaging; and international charities have lost trucks and drivers.

No one's life is safe in Uganda. Ironically, one of the most hazardous places is Mulago Hospital in Kampala, with no working plumbing and no sure supply of medicines. Drugs are known as nurses' gold, and they vanish quickly into "magendo"—the black market. "When you walk into this hospital, you walk into an open sewer," said one grim surgeon, despondent at the lack of anesthetics, syringes, gowns, gloves, blood. "I haven't operated in four weeks."

127

From left: Mike Wells (Aspect), Michel Folco (Gamma-Liaison)

Unorthodox layout shapes

Often eye-catching pages can be made when the photographs or illustrations are not made into the usual shapes. Arbitrary cropping, overlapping of pictures, or the kind of counterpoint juxtaposition used in the spread to the right can be strong attention-getting techniques. Note that the three examples of this tool shown on this and the next spread offer complex picture patterns but all consolidate the pictorial elements into simple large areas. On this spread, the pictures form a large rectangle; those overleaf make a large chevron and another large rectangle. The illustration areas on all three examples are quite complex, but the design structure is simple; the busyness has an easily grasped pattern.

Juan Rodriguez, the professional golfer, shows polar sides of his personality in this quick-cut mixture: the steely-eyed match player and the crowd-loving cutup. The composite portrait is appropriate since these moods change just as abruptly with Chi Chi himself.

THE CHANGE IN CHI CHI

Rodriguez turns serious, but the humor still pops up
By **DAVE ANDERSON**

Wherever he goes at a tournament, people surround him, waiting for a wisecrack, anticipating a laugh. At a recent tournament Chi Chi Rodriguez looked into a gathering outside the lockerroom to renew an acquaintance with someone he had met in Puerto Rico when he was a relatively unknown pro there. "I knew you," the man said proudly, "long before you had a big name."
"My name is not big," Chi Chi replied gently. "My name is Rodriguez."
Around him, others laughed. But he didn't. He hadn't meant to be funny. He had meant it seriously. But his image had provoked the laughter. After a decade of comedy, Juan (Chi Chi) Rodriguez is supposed to make you laugh.

"What'd you get home with, Chi Chi?"
"An 8-iron," he might have replied.
"An 8-iron!" the spectator would say, incredulous at his power.
"Yeah, I had cement for breakfast."

But that Chi Chi has vanished. There's a new Chi Chi on the tour now—the real Chi Chi. He was there all the time, but not many people bothered to look under the straw hat of this 130-pound strand of copper wire who hit 300-yard drives between jokes. Mature now at 38, his sensitivity is more obvious. So is his thoughtfulness, his loneliness, his concern for others . . . and his involvement in golf. Every so often, the old Chi Chi reappears. But he performs under restraint, as if the comedian in him had been awarded momentary approval by the philosopher in him. Occasionally, after a birdie, he will revert to twirling his putter as if it were a matador's cape. Or raising his arms in triumph. Or chatting with the spectators. But most of the time now he's more concerned with

John Newcomb

A nationally known golf teacher, Bob Toski, offers his opinion of the upcoming crop of promising young golfers. The severely framed series of faces lends a pulsing excitement to the display.

BOB TOSKI RATES THE COMING SUPERSTARS

The prime ingredients of future superstars, as I see them, are youth, physical ability, emotional drive and

... and the Snead era has gone on for the whole century, it seems.

...ose we're in the ...era right now, ...eat as Jack's record ...he always has had ...ers ... Arnold Palmer, ...ber, Gary Player, Lee ...now Tom Weiskopf. All ...ely be called superstars ...up of superstar candida ...Miller, 27; Ben Crensh ...Wadkins, 24; Jerry He ...Mahaffey, 26; Tom Watr ...Jones, 28; Tom Kite, 24 ...st Fezler, 24—in that or ...them are almost there ...Some of them probab ...make it. And there ar ...aven't named who mig ...uperstars. Hubert Gree ...been on a hot streak la

A desire to say something about the enduring beauty of this green at the Westchester Country Club led the editors of Golf Digest *to make a composite scene with pictures taken from the same point of view in four seasons of the year.*

Richard Beattie

Johnny Miller
Ben Crenshaw
Grier Jones
John Mahaffey
Jerry Heard
Tom Watson

Repetition

Repeating the shape of the same object or person can engage the reader. The covers on these pages illustrate the two main types of repetition: simple pattern and sequential, or time-lapse, description. The cover below *talks about determining which patients need treatment first; the cover on the opposite page features a story on the common causes of fainting.*

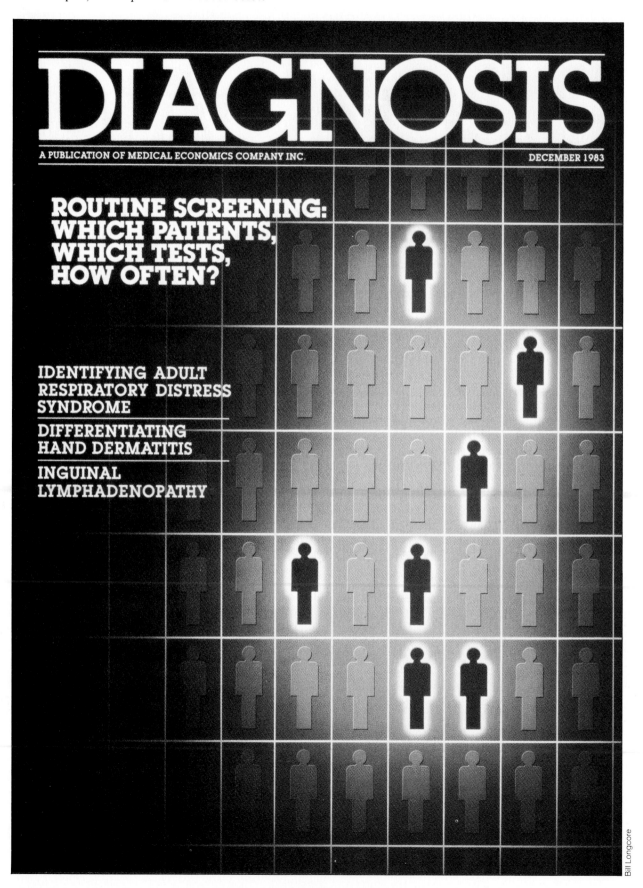

DIAGNOSIS

A PUBLICATION OF MEDICAL ECONOMICS COMPANY INC.

DECEMBER 1983

ROUTINE SCREENING:
WHICH PATIENTS,
WHICH TESTS,
HOW OFTEN?

IDENTIFYING ADULT
RESPIRATORY DISTRESS
SYNDROME

DIFFERENTIATING
HAND DERMATITIS

INGUINAL
LYMPHADENOPATHY

Bill Longcore

DIAGNOSIS

A PUBLICATION OF MEDICAL ECONOMICS COMPANY INC.   OCTOBER 1983

SYNCOPE: FINDING THE SOURCE

EARLY DIAGNOSIS AND
TRIAGE IN BACK INJURY

DYSPNEA: THE CARDIAC
AND PULMONARY FACTORS

DEMENTIA OR DEPRESSION?

THE UNDERLYING CAUSES
OF LEG ULCERS

Walter Wick

This series of drawings from Innovation *magazine follows the career of an ambitious young executive. Gradually he drains the color—and the authority—from his boss.*

John Newcomb

Elwood Smith's drawing for a poem entitled "Jimmy's Got a Goil" freezes four moments during which the musician lets his concentration slip.

Audience *magazine presented an article entitled "The Playmate Process" about Hugh Heffner's centerfold scouting methods. Robert Grossman's vivid illustrations, scattered among the text pages, spelled out the transformation from laundromat dowdiness to bird of paradise beauty, a fantasy.*

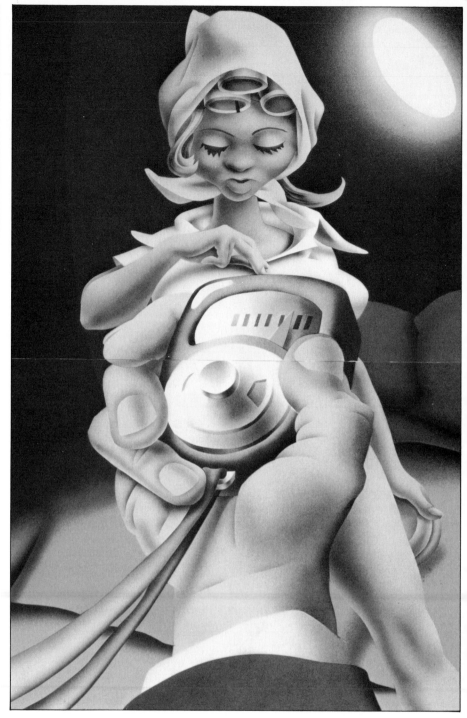

Robert Grossman

This full spread of repeating portraits of people laughing accompanied a story about Hollywood gag writing—making jokes for a living. Four artists collaborated on this illustration.

From left: Daniel Maffia. Philip Hays. Gil Stone. Seymour Chwast

141

Unexpected elements or dimensional mix

Graphic language can be made intentionally ambiguous to create surprise. In the illustration below, a photograph of coarse sandpaper was stripped into the sky area behind the figure of the golfer demonstrating sand play. On the opposite page the artist has placed three-dimensional objects over two-dimensional drawings for an offbeat illustration of the atmosphere in a tournament locker room.

James McQueen

with the other pros.

Like Bob Charles? Bob Charles is one of the tour's great putters. *No,* says Brown, surprising even himself with the line, *like Ray Charles.*

Every contestant's full name is stamped in green or orange plastic tape on his assigned locker. The lockers fan out from a small social area of thick tables and captain's chairs situated before a bar. Two white-jacketed attendants stand behind the bar doing little more than eavesdropping throughout the tournament, for there is no hard stuff available and the Hamms and Michelob beer dispensers are self-serving. On Wednesday one of the attendants hands Tony Jacklin a parcel that has been following him around from tournament site to tournament site for several months and now has finally caught up with him. It contains several pairs of pants and Jacklin pays the attendant the C.O.D. charges of $5.92. It is probably the last golf slacks that Jacklin, the eventual champion, will ever have to pay anything for.

Arnold Palmer, sitting with a few of the lesser lights on tour, calls to a burly, disheveled sportswriter, *Hey, Bob, with that haircut you look like Beethoven.* It is a weak, predictable crack and no one, including Arnie, breaks up over it. Palmer is biding time before he goes out to the practice area and then to the first tee for his practice round. For him as for the other players the locker room is useful as a decompression chamber between the worlds of career and family, and all its banal concerns, and the competitive life of the course. Even in the casual, anti-heroic air of the locker room, Palmer is special; when he moves, 20 or so others, mostly sportswriters, move with him.

An adjunct of the small social area is a table called the U.S. Mail Station & Message Board, the touring pro's lifeline to family, friends and business connections. The table is fixed with a divided metal rack where mail and messages to the players are arranged in alphabetical lumps. There are obvious

ead status into such a de-
vers like Palmer and Boros
ay from the U.S. Mail Sta-
Message Board with an
ading; others with nothing.
e thing happens with the
f scotch-taping messages
onts of individual lockers.
players will go all tourna-
ek without so much as a
UCK FROM DAD pasted
lockers. Nicklaus on any
y will have six or seven
varying degrees of conse-
JACK NICKLAUS, PLEASE
APH THIS PICTURE OF
JACK, CALL MEXICO CITY
OR 609 . . . PLEASE, JACK,
CTURE BRINGS BACK
ES . . . CALL DALLAS,
PERATOR #6 . . .

eryone sits around joking
g other players' messages,
on Wednesday. Ray Floyd
fully through the sociable
f his fellow pros in order
ne with a sandwich and a
r his locker, and glower
at the state of his game.
e Crampton, the solemn
n, goes directly to his lock-
inishing 18 holes and calls
platoon sergeant's voice,
Can you find some street
me, please!
f, after all, is a business.

Thursday, this businesslike
quality of pro golf is in fact
diminished by unusual con-
f play—high winds, some
a great deal of controversy
e suitability of the golf
self. *Why don't we just call*
j off, quips Larry Ziegler,
good-humored pro seated
f the large tables in front
ocker room bar, *and go*
e Astrodome?
ld beer on tap, or the vari-
rgy drinks like Gatorade
kkick that are available, do
a as attractive to the play-
ng in from a round as does
There is a buffet set out
y, under a sign reading
OR CONTESTANTS ONLY,
oming players stop there
e themselves dagwoods of
y; bologna or salami. First
down half or more of the

sandwich, as though absolutely
starved to death, and then they settle down with the drink of their
choice.

The pro golf locker room is odor-
less. The players don't come in off
the course, shed their clothes and
march around half-naked for an
hour emitting the sweat of their
work and then, as in a club locker
room, proceed to shower and purify
themselves with foot powder and
pomades and sprays. At most they
change from street shoes to golf
shoes and back again, and tuck in
their shirts. Yet their actions can
have an almost matadorial solemn-
ity.

The bad weather produces a fair
amount of sour grapes and bad
jokes.
How'd you stand at the turn?
Our best ball is 45, is all I know.

I just wondered if you got dis-
gusted out there, like I did?
Well, sure you do, but you still
got to hit the sonuvabitch.

Some of the younger pros, having
completed their rounds, sit in ease
at the big table. Periodically Dudley
Wysong sips from his cup of Michelob, and he gives the impression he
is listening to the banter, but since
he never shows any expression, finally you realize he is absorbed in

something else. One of the others,
Rives McBee, is asked when he is
going to leave. McBee, a good looking, genial sort, confesses, *I think*
I'm just going to wait here until
Nicklaus comes in. He's 43 going
out. And he's already said he
doesn't like the place. I've got to
see his face. Even inside the locker
room there are superstars.

Nicklaus finally arrives and, perhaps to McBee's consternation, fails
to overturn the buffet table in anger
at the day and his score (an
81). Instead he back-slaps Orville
Moody, one of his playing partners, and says, *Gosh, if he hadn't*
sunk that birdie putt on 18, I'd of
had a nice day. He pauses. *Orville,*
you don't mind my kidding, do you?
Moody, characteristically dead-
pan, says, *No, I don't mind.*

Nicklaus sees Knudson at the
table: *What'd you get, George?*
82, says Knudson.

Nicklaus repeats the figure, then
doubles over laughing earnestly,
and George begins to think it's
pretty funny, too.

Although a good part of the
day goes by without any
reference to it in the locker
room, Friday is the day of the cut.
The field, when Friday's scores are
in, will be chopped in half for the
play on the weekend and the shot
at the money. Despite the high spir-

its that went with the high scores
of the prevous day, everyone knows
how important a good score today
will be.

On a wall behind the buffet table
is a sign-up blackboard of the
club's saying HAZELTINE SENIOR
LEAGUE and below are the names
of members like Anderson, Owen
and Kelley. Someone has also
scrawled SAM SNEAD as a joke.

Snead himself does not acknowl-
edge this graffiti — any joke about
his age has got to be an old one,
he figures—and when he comes in
from his round, he draws half a cup
of beer and sits down. Shortly his
caddie approaches and in a reve-

"Coming in from
a round, a
player is more
hungry than
thirsty. He bolts
down half or
more of a
sandwich first,
as though
absolutely
starved to
death, and then
he settles down
with his
Gatorade or his
cold beer."

Robert Heindel

Another deliberate mixing of dimension is shown in
this Rolling Stone *illustration for a story entitled*
"Living single, sleeping double."

Lib Cummings

Daniel Kramer

Three brothers, highly successful taxidermists based in Seattle, were shown among some of their own trophy heads. The result was an amusing alternative to the usual boardroom portrait of top executives.

62   MEDICAL LABORATORY OBSERVER

The photographer's technical wizardry allowed these prescription drugs to explode from a computer screen. The pills even cast shadows on each other as they burst out of the scene.

Adjusting doses with a microcomputer

This hospital's automated drug monitoring system quickly does complex calculations and maps concentration-time plots. It also tracks clinicians' use of the data.

By John D. Meyer, Pharm.D.

Physicians at our hospital can now receive faster and better therapeutic drug monitoring consultations, thanks to a desk-top microcomputer system.

Several members of the pharmacy department helped develop this system, which currently embraces four aminoglycoside antibiotics: amikacin, gentamicin, kanamycin, and tobramycin. Plans have been made to extend it to bronchodilators, anticonvulsants, and cardiac drugs.

The computer performs sophisticated and individualized pharmacokinetic calculations, including drug half-life, volume of drug distribution, and suggested dosage adjustments. It issues graphic plots of serum concentration versus time, which help greatly in discussing cases with clinicians. It also quickly retrieves past drug data on readmitted patients. A separate program stores statistical information concerning the effec-

tiveness of therapeutic drug monitoring and the degree to which clinicians follow our dosage recommendations.

The project required a large investm[...]
spare [...]
relati[...]
Hard[...]
$2,0[...]
with [...]
disk [...]
printe[...]

We [...]
exper[...]
and i[...]
ated [...]
simil[...]
one [...]
know[...]
the c[...]
levels[...]
tholo[...]
shoul[...]

Pr[...]
doses[...]
This [...]
pape[...]
stora[...]
manu[...]

The author is assistant director of pharmacy/clinical and educational services at Hamot Medical Center, Erie, Pa.

Dimensional mixture is used skillfully on this cover featuring an article about finding the best-rounded individual to head a laboratory section staff.

MLO
MEDICAL LABORATORY OBSERVER • JUNE 1984

SUPERVISOR SELECTION: HOW TO PICK A WINNER

Guidelines for laboratory administration—
A new series

Using work station balancing to increase productivity

Marketing a new product from your lab

A training and in-service program for the nontechnical staff

Stephen E. Munz (art by Barbara Slocum)

Visual puns

Probably the strongest attention getter in the designer's armory is the concept visual, the graphic pun. Shown below is the careless doctor who incurs big legal trouble by neglecting small errors on a patient's progress chart.

Stephen E. Munz

This article described the harassment of a physician by a psychotic patient who was a walking time bomb.

Stephen E. Munz (art by Kathy Jeffers [head] and Margaret Cusack [body])

The U.S. Army modified the M-14 assault rifle until it was unreliable; the designer had a replica gun designed to reflect this.

Stephen E. Munz (art by Janis Conklin)

Stephen E. Munz

A doctor tells of unsympathetic treatment of his family by an unfeeling colleague. A brick pattern was projected onto the model's white coat to create an apt portrait of the callous physician.

The problem was to dramatize a list of checkpoints for the manager to use while interviewing job applicants. Advice was given to help the interviewer pick the right person for the right job more often.

A story from Psychology Today *explored the medicinal uses of marijuana.*

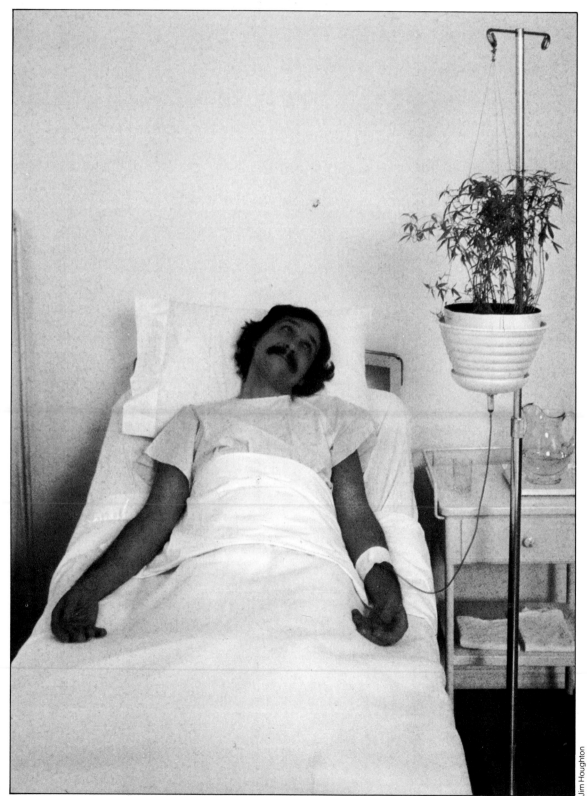

Jim Houghton

The editor's advice: use the lightest golf clubs you can find. The feather was carefully crafted from a headless driver and a painted piece of Styrofoam. The golfer is pro Frank Beard.

New tax laws gave birth to this super sundae garnished with 1040 Forms. The title of the story was "The IRS serves up more fiscal indigestion."

GOLF DIGEST

75¢ 14117
December 1974
Largest circulation of any golfing publication

THE CASE FOR LIGHTER CLUBS

HOW TOUGH SHOULD TOUR COURSES BE?

Frank Beard

Medical Economics offers its readers "Six practice-building strategies that really pull in patients." The doctor's office becomes a large magnet.

Ken Schroers (art by Asdur Takakjian)

Jerry Cosgrove

Jerry Cosgrove

A battle raged between devotees of golf balls covered with the traditional rubber balata material (which gives more shot control) and evangelists of the new plastic Surlyn-covered balls (longer lasting, harder to cut). The struggle of the partisans was represented by a fight scene with the balls themselves as adversaries. The costuming included red vinyl boxing gloves and satin trunks whose stitching read "Slugger Surlyn" and "Kid Balata."

An alligator gar, the southern gamefish that you have to shoot to kill, is shown with built-in heavy armor.

Stephen E. Munz (art by Janis Conklin)

"Plant a profit in the good earth" the story advises. An ear of corn is used to symbolize the wisdom of investing in farmland and annual cash crops.

The story was about a man who believed he could save money by functioning as his own contractor during construction of his house. The concrete foundation developed a large and seemingly unfixable crack shortly after it was poured. The rest of the project was a losing battle against one construction mishap after another. The dream house began life as a lemon.

159

*Superb mechanical
control by the photog-
rapher produced this
vivid symbol of the
ultimate dirty trick.*

Angus Forbes

*The national obsession
with big-league baseball
gave birth to this image
of the true afficionado.*

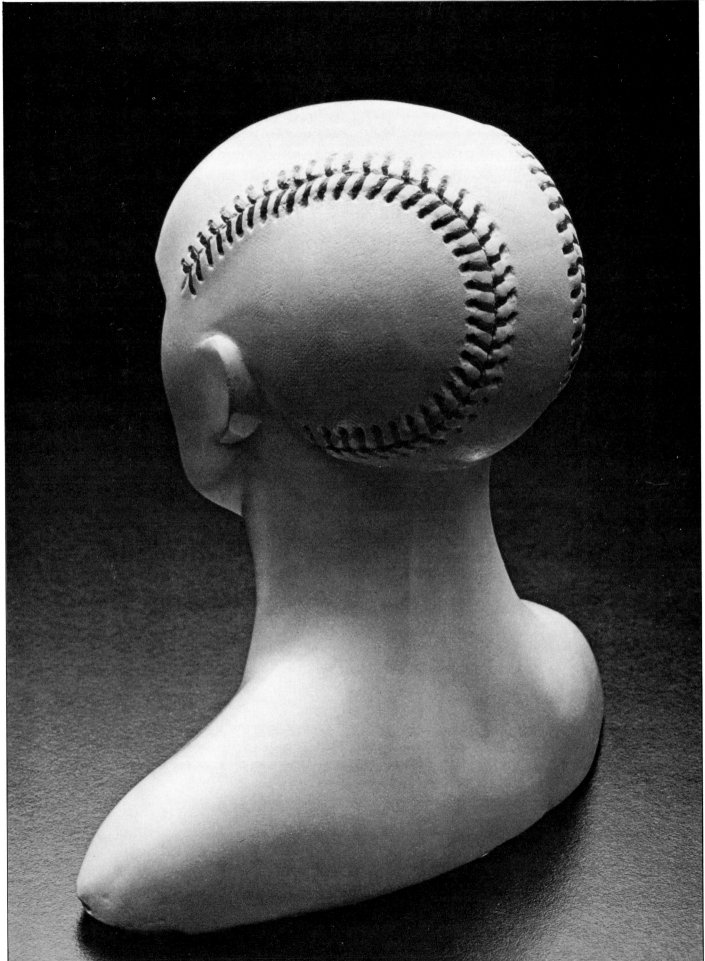

from the rest of the thought, the rhythm of the title was destroyed, and the words chosen for emphasis had little power to provoke curiosity. It is difficult to believe this use of very large type would stop a reader in a positive way; though it may catch his eye for a moment, it would do nothing to make him read the story.

An illustration in this chapter shows a headline that says "HELP! I'm a financial disaster." It could also have read "Help! I'm a FINANCIAL DISASTER." Another spread has a headline that refers to an aging golfer who was once great; the most surprising element in the statement was emphasized in one word: "He won the U.S. Open? TWICE?" You must be careful with the words you choose to blare forth. Make those verbal blasts count—and don't use this tool too often. Two or three places in one issue is more than enough.

6. Subject focus on the reader. The dramatic quality of any lead picture will be stronger if the subject is looking or pointing directly at the reader. This is particularly effective when the picture accompanies a headline that either says something about the person shown or states something the person is supposed to be saying. If the opening photograph portrays a proto-typical farmer in overalls leaning on a pitchfork, for instance, and the headline reads, "I can only take one more year of this," the combination is used well. The subject, the disgruntled farmer, makes eye contact with the reader and speaks directly to him. As in daily life, the most effective way to talk to someone is to make eye contact.

7. Unusual point of view. The fascination of images in this category is that of seeing familiar objects or people from unfamiliar angles. If you can direct your photographer to get above or below the subject you can inject freshness into the image. If you can make the picture look as though it were taken from a place where logic indicates there was no room for the camera, you can add to the picture's appeal. The most intriguing shots are those that make the reader wonder how on earth the pictures were taken, the impossible-to-get photos. One of the most effective images of this sort was done for the opening page of a special section of putting tips in *Golf Digest*. I asked the photographer to shoot a golf ball as it was falling into the hole—but I wanted him to shoot upward at the ball as though the camera were at the bottom of the hole. His solution was fairly simple. He cut a round hole of the correct diameter in a piece of black paper, carefully arranged a fringe of short turf around the edge, taped a golf ball so that it hovered on the rim of the hole and looked as though it had to fall. Then he shot from below the black paper to show the hole, the ball, and a blue sky with clouds; the shot was convincing.

8. Emphasis on telling detail. Very often the heart of a visual story lies in a small detail, a part of a scene or a tiny side act happening on the edge of the main show. Frequently a picture story will gain tremendous impact and poignancy if the designer gives importance to one of these small but evocative details.

The two most vivid examples I've seen of the use of this tool both appeared in *Life* magazine. One was in a pictorial record of a day in a big city hospital emergency room, and the other was part of a short story about famine in Biafra. In the emergency room story, the largest picture in the set was not the obligatory group shot of doctors and nurses working frantically over a car accident victim but a tight shot of the feet of the doctors and part of the floor. The feet were covered with surgeons' plastic slippers, the floor showed pools and puddles of blood as well as discarded tubing and bits of paper, and everything—shoes, trousers, and refuse—was splattered with blood. Nothing could have spoken more eloquently than this picture with its jarring mix of the bizarre and the routine to represent life in a crisis tank, the trauma center of a busy hospital.

The Biafra feature reached an emotional climax with a large photo of a life-size human hand—a Caucasian hand—holding a small, shriveled hand that looks like a monkey's paw. When we read the caption we discover that the tiny brown hand belongs to a starving ten-year-old child. No picture of haunting young eyes or swollen bellies—the usual ways to describe hunger—could have been more affecting.

9. Unorthodox layout shapes.

Nowhere is it written that pictures in a magazine must be square or rectangular. If a feature gains character and surprise by doing so, crop photographs into narrow strips or fit them into irregular, interlocking shapes within a grid of lines that resembles building blocks. Pictures may be contained within the areas of other pictures or may be used together in a quick-cut-now-and-then sequence— a counterpoint of two types of photos, one taken, say, when the subject was young and one taken recently that shows him aging.

I recall using this treatment to show two different aspects of the golfer Chi Chi Rodriquez. He was an irrepressible clown between shots on the course, and his antics always delighted the spectators. But there was another side to Rodriquez: the steely-eyed competitor. A large closeup of his face showed him frowning in concentration as he lined up a putt. Into this shot several small photos of him dancing around the green and waving his putter were inserted. The combined photograph was a more efficient statement of his multi-faceted personality, and the graphic was a stopper.

10. Repetition.

That print design shares some characteristics with music is demonstrated by this graphic tool: an object that is repeated several times attains a musical rhythm to which most people respond as they would to a catchy tune. The drumbeat of repetition in a layout can furnish a strong visual element to get the reader's attention.

Two types of repetition can be used. The first is a simple repeating pattern of similar items on the page; the second is time-lapse repetition, the reappearance of a key factor in a sequence, a chain of pictures showing the same object as it changes.

11. Unexpected elements or dimensional mix.

This tool is exemplified by the illustration that shows a two-dimensional subject interacting with a three-dimensional object—a fully modeled painting of a girl lying on the grass next to an outline drawing of a boy, for instance. There are many fine combinations using this dimensional confusion, the interplay of flat and round.

Another pictorial breach of logic in a visual might be represented by the treatment once given an instruction story about sand play for *Golf Digest*. Bob Toski, the nationally known golf teacher, wanted to demonstrate some differences of stance and swing path that applied only if the golfer found himself in sand. The teaching points were represented with full-color realistic drawings of Toski demonstrating the recommended stances or swings. The feature was given added graphic appeal by photographing coarse sandpaper and having the printer strip the sand image into the areas of the drawings where sky would show. The attention-getting result showed the golf teacher giving a series of lessons in front of a curtain of sand.

12. Visual puns.

Last, but certainly not least, we have the symbolic statement, the pictorial paraphrase that carries its own message. This is the designer's most radical way of getting the reader's attention, his or her poetic parallel for the author's words. This device can inject a healthy amount of entertainment in the graphics of a magazine—witness the spreads shown in this chapter: the flawed, always-in-need-of-repair house shown as a lemon with doors and windows; or the careless physician making notes on a patient's progress chart while his head is obscured by clouds. These are visual puns.

All of these basic graphic tools are legitimate ways to create peaks of excitement and surprise the reader. However, having discussed the high points of pacing in a magazine, a few words on overall design character, the setting for these emotional jewels, may be in order.

How an effective magazine should "feel" is subjective and open to debate; proponents of the so-called "New Wave" style of editorial design would object to my criteria. For designers of this school the treatment of subject matter emphasizes graphic techniques: random pink, aqua, and blue violet amoeba shapes swim behind text as funky headline type wanders freely in and out of photos or illustrations. The whole graphic language smacks of graffiti and iconoclasm, and the goal seems to be to fill the reader's senses with incidental noise, to dazzle him with a blizzard of unstructured elements. Such frenetic page design does more to distract than it does to invite, instruct, or inform; it seems more concerned with *how* something is said than *what* is being said, and, as such, it doesn't match my standards of good communication.

There is confusion, even among professional designers, about the term design as it relates to magazines and books. A well-designed format is one in which the writers and artists who contribute their work and the editors who harmonize the group of messages it contains may speak freely and naturally with the readers in their target audience. The normal arrangement of text and illustration should have adequate individuality, consistency of look from issue to issue and an easy familiarity.

Illustrator/designer Milton Glaser says, "People don't understand what design is. It's finding an appropriate form for what you want to express. Some magazines can be very flashy and daring and interesting visually and find that has no utility as far as the reader is concerned. Others can be mostly visual and work, like *Architectural Digest* and *House and Garden*, which are dazzling. Sure, design very often serves as a sort of suit to cover up a flaw in somebody's appearance, but those are the cases when you always know what's going on underneath the suit anyway."*

A magazine should be analyzed for its body language, for the overall impression a reader gets as he first thumbs through an issue. The way a person arranges his body speaks volumes about how he really feels; he may signify resentment by folding his arms and crossing his legs, or he may actually turn partially away from a partner in conversation. He may fidget or grip the chair arms so tightly his knuckles turn white. So it is with magazines.

As you flip through an issue of any magazine at hand, ask yourself these questions: Does the graphic tone of the pages radiate calm and confidence, vigor and purpose, or does it seem strident, fitful, and insecure about the information the articles are supposed to carry? Are the pages an inviting setting for a few well-chosen, really different graphic displays and curiosity-provoking typographic elements, or are you faced with a cacophony of competing accents that leaves you doubting where to start reading? Potential readers sense this self-confidence or the lack of it in a magazine's design—and so do prospective advertising buyers, the media decision-makers in agencies who quickly gauge the publication's value as a vehicle for their ads. It is important to control the body language of a magazine.

Earlier in the chapter we spoke of the desirability of

Adweek, March 1984: 16.

making the magazine a friend to the reader, giving it a recognizable voice that entertains, advises, informs—a familiar voice that the reader comes to trust and welcome. To advocate such a tone in a magazine is not, of course, to suggest that a publication be dull or safe. Far from it. The real goal is to make every issue a lively mixture of the expected—the basic typographic format and the promised coverage of subjects of value to the reader—and the unexpected—the specific graphics fireworks. The high points, the emotional splashes, must be set like gems in less competitive material that has a conservative, uniform character. The editorial typography should be natural, accented sparingly with bold elements, and should be generally unobtrusive. This forms the counterpoint melody for the thrills and trumpets of visual peaks, the graphic big guns of the issue. A magazine's normal typography is the editorial mortar that holds the book together within one issue and from one issue to the next. This tone is what builds a personal relationship with the reader; it signals that he is seeing this editorial friend whose company he has come to enjoy. A human friend will look much the same every time you see him or her, and this is reassuring, but you like to see them because you like talking with them. A good magazine works in just the same way.

So, obviously, the use of a continuing format, a repeating style for designing pages, is highly recommended. The format should have personality but should not chop and torture the material as it moves from editor to reader. Eccentric, hard-to-read text faces and ever-changing headline styles are the wrong outlets for a designer's energy; the effort is better spent on creating entertaining and meaningful illustration—photographs, paintings, drawings, and charts—that dramatize the content of the editor's articles. The type format should allow the editor's message to be carried to the reader with the least amount of friction in transit. The ease and speed with which the thoughts reach the reader ought to be maximized with typography that is unself-conscious, not mannered.

And, of course, a repeating format is indispensable for a magazine that includes any substantial amount of advertising. A reader should not have the slightest doubt when he looks at any page in the issue—it's either yours (editorial) or theirs (advertising). This is highly desirable not only for the people who produce the magazine but for the clients who advertise. Advertisers want their messages to stand out from surrounding material so that they may tell their separate stories with clarity and strength also.

These, then, are the ingredients of a good magazine. If you make regular, thoughtful use of the basic graphic tools for creating surprise and variety and if you give enough attention to the overall "voice" of your publication, your magazine will probably grow and you will become a better graphic conversationalist.

Stan Mack

TRAIN YOURSELF TO BE AN EFFECTIVE DESIGN THINKER

Stan Mack's stumped designer doesn't surrender to his block; he simply puts on his designer's head. (Figuratively speaking, this is valid, for creative design requires a voluntary change of mental approach.)

Having worked as a graphic designer for some twenty years, I had come to grips with the process of getting visual ideas through my own daily activity. While researching and preparing my presentation for the *Folio* seminar, however, I felt the need to talk to others who make their living dramatizing words—other top professional idea-makers working in the applied graphic arts. I suspected that the creative process, work sequence, and personal problem-solving technique of each of these successful performers would reveal enough similarities and enough pattern to be of use to the beginner. I believed that insight from these interviews would harmonize with my own and that elements common to most or all of them would yield points of concrete advice for the designer who wished to improve his or her craft.

The interviews were extremely successful, and there is, indeed, much in what these accomplished idea-makers say that can be of help. The information was of two kinds. First, there was discussion of their individual methods for attacking a communication problem—the environment they preferred, whether they worked best alone or by playing mental ping-pong with others, whether they had to shift mental gears to think about visual solutions, how they coped with blocks, and so forth. These were the daily conditions of the work.

Second, there were personal observations on useful habits and disciplines, the best education and conditioning for performance in this line of work—suggestions as to how a young designer might set an individualized curriculum and train his or her mind to develop the skills needed to be a working graphics professional.

The group of high-level idea-getters chosen for this book does not comprise just art directors, although some of them are or have been. Rather, the group represents a deliberate mix of art directors (both advertising and editorial), illustrators, cartoonists, copywriters, painters, and graphics teachers. All of them addressed the problems of creativity, each from his or her own view, with perception based on long experience.

The group of interviewees, referred to here as "the panel," included these top performers:

George Lois of Lois Pitts Gershon, Inc. A seminal figure in American advertising and graphic communication, he started his career with the legendary Bill Bernach, to whom words, pictures, and the human needs of consumers were all entangled. (Bernbach is credited with the idea of using the creative team approach to problem-solving. He combined artists who could make copy suggestions with copywriters who could give graphic input and thus tore down the traditional wall between picture and word specialists. He trained a dynamic new generation of ad builders.) Of course, Lois has created campaigns for hundreds of products, places, and people in the years since Doyle Dane Bernbach. In fact he has helped found four pace-setting agencies of his own, each made to escape the complacency of the previous one. Not least among his graphic world-shakers were the ninety-two idea covers he created for *Esquire* magazine starting in 1962. Strong message, wit, and audacity mark his work.

Milton Glaser. A multifaceted graphic genius, he and the innovative Push Pin Studio group changed American illustration history with their uninhibited, fresh graphics and their naive/sophisticated drawing styles. Glaser is certainly one of the best known and most widely respected figures in professional art, both in this country and overseas. Superb draftsman, first-rate designer, and idea-maker par exellence—he has it all.

Seymour Chwast. Another of the founders of Push Pin Studio, Chwast has occupied the top rungs of American and international illustration since the early 1960s. His ingenuous style, more than that of any other artist today, identifies a Push Pin school of visual interpretation that furnishes much of the best graphic communication done today.

Elwood Smith. Smith is an art director turned illustrator whose hallmark has become a warm, folksy kind of humorous drawing that owes much to one of his early heroes, George Herriman, the creator of the comic strip, *Krazy Kat*. Smith works in a refurbished house in picturesque Rhinebeck, New York. He lives there with his family, his pets, his music, and his "preference for working with ideas—not just making drawings."

Guy Billout. A French illustrator who now works in New York, Billout is much in demand by national publications needing his evocative, strongly patterned drawings of specific objects that illuminate the general.

His is a quickly grasped, universal language, not only because the rendering is clear and simple, but because the conceptual quality of his graphic statements is so high. The unique way he delivers the message distinguishes his work from that of imitators and keeps his telephone ringing.

Anthony Angotti of Ammirati & Puris, Inc. Creator of the highly entertaining Club Med television commercials ("At Club Med you can exercise everything . . . including your right not to exercise anything"), Angotti was named to *Adweek* magazine's Creative All-star Team of 1982. Witty, warm, and articulate, he is the very model of the creative heavyweight.

Carveth Hilton Kramer. Former art director of *Psychology Today*, he is a mass producer of striking editorial graphics, a fountainhead of visual ideas. For the years he generated the graphics of *PT*, the magazine was regularly represented in top national award shows.

Dik Browne, creator of the comic strip *Hagar the Horrible* and collaborator (with Mort Walker) on the comic strip *Hi and Lois*. Browne was an art director and commercial illustrator; he designed the symbol for Birdseye frozen foods and drew the trademark for the Campbell Soup Twins. He was chosen for this panel because of the communicative power of his humor, because he is eloquent about ways to make ideas come, and because he must create twenty to thirty ideas a week forever!

Stan Mack. Mack has functioned as an art director (at the *New York Times* Sunday magazine) and as a humorous illustrator for national campaigns and publications, but his most significant recent work has been using an art form he invented. He executes comic strips that reflect modern life and working; their special device is dialogue that is guaranteed to have been overheard—what the characters say was really said and the situations are taken from life. These little real-life dramas, presented with humor and sympathy, run regularly in the newspaper *The Village Voice* and in *Adweek* magazine. He also teaches design at the School of Visual Arts in New York City.

Bruce Bacon. Painter, photographer, and design instructor at the School of Visual Arts in New York City, Bacon continues to be an avid student of creative theory himself. His literate, inventive paintings are circulating to clients all over the world.

Nancy Rice and Tom McElligott. Rice, an art director, and McElligott, a creative director, have collaborated to create some of the most arresting ads and some of the best-integrated verbal/visual pieces of communication to enter the national graphics shows in many years. Their ideas are characterized by strong and colorful language. With only a few months' work to enter at the New York Advertising Club Show in 1982, the fledgling Minneapolis agency, Fallon McElligott Rice, walked away with three Andy awards. In the same year the agency also got five Clio awards; they also received a gold and a silver medal at the New York Copywriters Club and completely dominated the local design show in Minneapolis and *Advertising Age* designated them the Agency of the Year for 1983—hot stuff, indeed. Their ideas capture the viewer immediately with verbal and visual wit and daring. These people are drawn to solutions that are conceptual and unorthodox; in fact, one of their first policies when the agency opened was to seek out clients that were willing to take chances.

Paula Scher. The former art director of CBS Records, Scher teaches a senior portfolio design class at the School of Visual Arts in New York City. Recipient of countless awards and graphics honors, she is much respected in top creative circles.

This, then, is my panel of consultants. Let's see what they have in common when they go to work on a problem.

One of the most striking observations had to do with "a favorite place to work." The expectation was that each would describe a favored retreat—a quiet study in a nice old brownstone, a sunny loft in Greenwich Village, a workroom that looks out on dense woods, or a cluttered nook in the back of an old farmhouse. I expected to hear of at least one sanctum each, a familiar environment where the mind would signal that it was time to get serious about making pictures.

Nothing could have been more mistaken. The more I asked, the more I was told that there was no fixed or official place in which to create. The only real studio is in the designer's head, and the furniture of the mind is the only real constant. Some of the pros actively spread their conscious working time over the landscape. Angotti says that some of his best ideas have come while stuck in a traffic jam on the FDR Drive coming into Manhattan. Stan Mack feels most at home, most creatively active, on the sidewalk with his notepad ("The final form is given the drawing at a drawing board in a room, but that's not the important part"). Glaser says he gets ideas or solutions while eating breakfast or riding in a taxi as often as he does in the "proper places"—in the studio or the conference room. The point is that graphic design is an occupation that permeates most other areas of life. The synthesizing portion of an artist's mind seems to work more or less continuously, even when the person is sleeping! Solutions to given problems have a habit of popping into view at odd times—usually when the body is relaxed.

Several of the panelists thought their creative productivity was highest at both ends of the day; the best ideas came to them as they lay awake for a few minutes after a night's rest and also as they lay waiting to fall asleep at night. These would be times of maximum relaxation—dark, quiet, with no images except the ones projected on the screen of the mind.

From the comments the panelists offered, a general problem attack sequence could be distilled to describe the progression of stages in solving a design problem. The nature of the items and their sequence, with slight variations, were common to all the interviewees.

1. The research stage. The problem must be defined and the communication goal must come into focus. The designer must absorb all available facts, either by reading the relevant material or by discussing the subject with the writer. At this point use of the Bite System can help to sift out the valuable parts of the message; it can help the designer understand the statement or product more fully, understand its essential nature.

2. The analysis stage. Here the designer begins expressing opinions about the subject, starts examining different aspects of the problem. In the Bite System, this level of exploration would consist of making source statements, the thoughts that form the basis for substatements.

3. The synthetic stage. At this point the designer will be making new combinations of elements, using free association and the substatements to play word games, phrase games, and fact games.

4. The first visual solutions. A number of pictorial paraphrases will emerge in crude sketch form.

5. Time away from the problem. The wise creative person will shelve these visual ideas and allow his subconscious mind to continue working on the problem, overnight if possible.

6. Review and supplement of the ideas. At this stage, typically, another way or two of solving the problem will result from the work of the mind's night shift, and these ideas, along with the ones created at stage four, will be refined—examined in the cold light of day to see how much vitality and punch they retain.

7. Testing the best idea(s). Finally, the most interesting idea or ideas are exposed to other people. The most effective will be obvious. Tom McElligott tells of the process for testing the appeal of embryo ideas at his agency: "Normally I will scrawl two hundred to three hundred copy versions on as many sheets of yellow scratch paper. Nancy [Rice] will read these and rough-sketch a picture that might accompany the headline for each version she likes. Then we pin up a selection around the walls of our conference room. We bring in several kinds of people—salesmen, secretaries, other art directors, or copywriters—and the idea that produces the most surprised laughter is usually the best. We know we're almost home when people start pointing and giggling."

For the designer beginning the process of solving a visual problem, there are specific pieces of advice that seem valid in the light of my experience and that of the panelists:

Adjust your environment to fit the activity. You must literally be able to hear yourself think. When you decide it's time to really face the problem, you should find a private place where you won't be interrupted for as much time as you think you may need for the first work session. McElligott describes a small, semisecret workroom in the back of the agency: "There are no windows and no phones. When I use this room, I give myself a generous slice of time and all callers are told I'm in conference." Seymour Chwast says he works in a busy design office with twenty other bodies in close proximity—but he admits that there's a "small, cluttered room" into which he can retreat with a drawing that needs more thought.

Virtually all of the interviewees said they began the creative process alone rather than in tandem with others. Guy Billout had recently created a set of illustrations for a *Time* magazine article on the quality of education in this country. The concepts were generated in a Rand Corporation-style think tank, a conclave of editors, researchers, art directors, and the illustrator. Undeniably the ideas were good, but Billout expressed a preference for starting the creative process alone; he felt that this was more efficient and productive for him than the group effort, at least in the first stages.

My own experience reinforces this impression. In our own shop, manuscripts for articles are distributed a few days before the art conference with editors; by the time we meet, each designer and each editor will have done his or her own creative spadework and we will be able to put our best ideas on the table. There is no denying that ideas can be improved or that new hybrid ideas can appear when you can tug the best ones back and forth with other creative people. But start alone.

Relax. Do what you can to put aside other problems you may have. If it helps to gentle yourself down with your own voice, do so; talk to the different parts of your body, ask them to relax separately. You may feel a little foolish at first, but it has been proven that this kind of biofeedback can lower your blood pressure and put your mind in a more receptive mode. Carveth Kramer equates art planning with many kinds of sports: "The best results come if you don't try to force them—the golfer with a relaxed, easy swing will usually hit farther than the clamp-jawed hacker who lunges at the ball with every available muscle." It is significant that many of the panelists noted that ideas seemed to come just at the point where the mind was drifting between the waking, conscious state and the sleeping, unconscious state—that is, when it had just awakened and just before it fell asleep. So, do what you can to relax before you start work. You may even wish to put a little quiet classical music on the tape deck; some of the panel members used this. As long as the music you play

doesn't demand your intellectual involvement—say, the lyrics of a Broadway play or the words of a pop ballad—and doesn't disrupt your conversations with yourself, its effect is likely to be beneficial.

Be optimistic. Have faith that you can make good visual solutions and that usable ideas will come to you. This is terribly important. Many young designers beat themselves before they even begin by concentrating on their fears, by dwelling on the possibility of failure to the point where failure is all they can think about. They convince themselves that they are blocked and will remain blocked, for ever and ever, amen. For a creative person this is a real enemy: self-doubt. Glaser talks about knowing the ideas will come. Even if some are complete in one morning and others take a couple of days, time spent thinking about a problem is time invested—not time wasted. McElligott speaks of "believing I can get an idea," and Kramer says he has faith in his special skill, his ability to make new things from parts of old. Open your mind to possibilities. Absorb all the facts of the problem through discussion or reading before you begin the creative work, then start your exploration in a spirit of play. Be as whimsical as you like.

It is important to branch out with your speculation at the very beginning, to explore several paths of inquiry early in your thinking about a problem. Psychologists tell us the human race is divided into two general categories of thinkers: convergent thinkers and divergent thinkers. The convergent thinker begins narrowing the logical possibilities from the start; he is convinced that there is only one true answer to any question—a Holy Grail solution that can be found by a ruthless process of elimination. Such a thinker will stake everything on the truth and appropriateness of one final idea. The divergent thinker, on the other hand, senses that there are many good answers to a given problem; poking and probing in many places from the start, he explores as many different lines of logic—and even illogic—as he can.

It is obvious which type of thinking must be cultivated for use in creative work. Like many other habits that can be controlled or changed, your thinking can become divergent if you put your mind to it. You can stop being a convergent thinker or at least guard against this tendency. Bruce Bacon speaks of planning for happy accidents and widening the designer's field of vision; that is, he suggests adding the dimension of time to what is seen. He says, "Stretch the area you see—look beyond, before, and after. Orient to expand your field of vision." Stan Mack describes using "peripheral vision, the random word or doodle seen out of the corner of the eye." These are other ways of advising the designer to avoid limiting the imagination early in the creative process.

Shift mental gears. Most of the panel members reported a conscious, voluntary change of attitude at the start of a problem-solving session. Basically, the change had to do with becoming less critical than they might be while conducting other affairs of life, say, crossing a street or buying a gift. Such advice could be expressed this way: put aside your inner censor, the little voice that whispers, "Don't make a fool of yourself." Don't worry about what your spouse, your mother-in-law, or your boss might say. No doodle or expression is too absurd. Every impulse should gain acceptance at this point; every thought should be a "why not," not a "why." Angotti makes an effort to indulge his fantasies while thinking about a problem: "I like to play 'what if' games—what if I could do this with the product, what would I like to do or see that includes this thing?" There's no room for self-criticism here.

Reach for illogic. As you make combinations that result in substatements and their visual equivalents,

try for mismatches and unusual or unexpected pairs of elements. Tony Angotti mentions the frequent use of this approach; he chooses opposites—say, a tennis player and a bulldozer—and imagines heavy-duty hydraulic shock absorbers on the person and sneakers on the machine. The product is lifted out of its expected context, and the new combination becomes vivid analogy. Practice making combinations that would never really fit together: awkward marriages of objects and functions may make valuable statements about your product or subject precisely because of their outlandishness or exaggeration.

Violate natural law whenever possible. Reverse the principles of physics and make hard things soft, soft things hard—bend a bone, create a stairway of water, suspend a bridge in midair, have a man lift a corner of a building with his hands, let trees walk. . . .

Use starter images. Several of the panelists reported using visual aids or thoughts that helped put them in the proper frame of mind for creativity and helped them focus on the problem. George Lois talks of his: "After I've taken in all there is to know about the problem, I sit back, relax, and swill the thing in my mind. I just let it wash around inside my head, much as I would roll a wine around in my mouth so that I could determine its true flavor. As I swill the information I begin to *feel* the knowledge of the subject, not just *know* it. The ideas begin coming, and they appear as fruit on a tree before me. I must be alert and quick enough to pluck the ones I think I want while the tree is still there. I think of ideas—solutions—as fruit that I reach up and pluck." Stan Mack describes a similar device as he talks about "indulging the mind, letting the mind paw freely over a problem, letting it wool it around at will."

Carveth Kramer induces a constructive schizophrenia in himself to explore a problem—really a case of self-inflicted multiple personality. First, there's the Conscious, the part of him that explains the problem and functions as a library for further reference during the ensuing conversations. Second, there's the Subconscious, the one that elaborates and offers solutions, the one that makes strange and unexpected combinations. Third, there's the Friend, the kind of no-nonsense, familiar critic you might find in a spouse or a grown child—the one who keeps hold of your string as you rise in a trance of creative excitement, the one who says, "I've seen it before" or "I don't get it." Finally, there's the Mediator, who merges all concerns and reconciles all disagreements. After these four personae hash out the problem and dissect its parts, they reach consensus. Kramer says he trusts them all and does nothing to muzzle any one or argue for one at the expense of the others; he maintains his role as a manager who has assigned his able staff a task. It is a distinctive approach, but it seems healthy and productive.

Dik Browne thinks of his typical reader as a foot and the reader's life as a shoe; "In those places the shoe doesn't quite fit, where it makes blisters and calluses—there is where I look for my ideas. The average person feels discomfort at many points recognizable to us all—frustration with a job, annoyance with a class of professionals (plumbers, taxi drivers, lawyers), impatience with the routine demands of family and friends, the amount of energy and time needed to fill our places in society, household chores—there are plenty of spots where the shoe rubs."

Stan Mack targets the same kind of thing when he looks for areas of abrasion, where something said or done is slightly upsetting to him. He finds this sort of funny/sad ingredient in such situations as two married people who share a tiny apartment; she's expecting and he wants a basketball player son but worries about there being enough practice room in the apartment. Or he finds it in the complaint of a freelance artist who swears he'll never work again for "a middlewoman," a female art director whose approval of his work is reversed by a male creative supervisor.

Elwood Smith sees himself as a large creative storage battery. He says he frequently does all the necessary research for a problem and then deliberately procrastinates for half a day. He feels that his subconscious mind is working on the problem, storing up more and more energy, charging itself with more and more creative electricity. When he finally sits at his drawing board, the energy is released in bursts of insight, and the ideas usually tumble out head over heels. Attach the wires (study the problem), charge the battery (let the subconscious work on it), release the energy (doodle out the ideas); this is the work image he has.

You, the reader, will have to make your own starter image to fit yourself. Very probably none of the ones described above would meet your needs, though you may wish to try one or more and see if it helps.

Keep the human connection. Never forget that, whatever you say, the message is meant for human beings. When Browne and Mack talk about finding places in the human experience that show need, they are aiming for spots of significance in the lives of us all. The object of your labor, the person to whom your editor or copywriter talks, has personal desires that have been the same throughout recorded history—desires for security, for respect, for leisure, for love. It is the communicator's job to use these desires to make the viewer respond, to make him feel and do certain things. As a designer, you must not lose sight of the human animal. Angotti says, "When I start exploring a problem I isolate the benefits of the product. Personal use is the key—that special relationship I can highlight for the consumer and the product. I use the human condition as a conduit."

(Text continued on page 228)

IDEA MAKERS AND THEIR PRODUCTS

The following pages show the work of top professionals interviewed for this book. These examples furnish clear standards of quality and effectiveness for communicators of all types. The solutions are timeless because the messages are strong, vivid, and human.

GEORGE LOIS

It's difficult to overestimate the influence this man has on communication of all sorts but his area of greatest impact has been on the graphic designer's own daring. Before Lois became the enfant terrible *of Madison Avenue, most print communication was bland, preachy, or soporific. Over the years, oblique praise by his critics has resulted in such sobriquets as "the dull-be-damned art director" and "the elder statesman of graphic outrage." The fact remains, though, that he taught us to push to the very edge of taste in order to say something that matters in a way that will be remembered. Perfect examples of Lois's visual daring are the four classic* Esquire *covers shown at right. His*

design fee for each cover, $1000, was quite high for the time, but all of the proceeds were donated to an orphanage in Greece (Lois is proud of his heritage). The newsstand sales of most of the ninety-two issues on which he worked climbed over the million mark.

To the amazement of his colleagues, Lois still prefers to do his own storyboard sketches and to trace all his own headline type. His office is large and comparatively bare. An informal person in a formal atmosphere, he likes to work dressed in denims and old shirts or sweaters. Of course he keeps suits and ties in an office closet for important meetings or lunches.

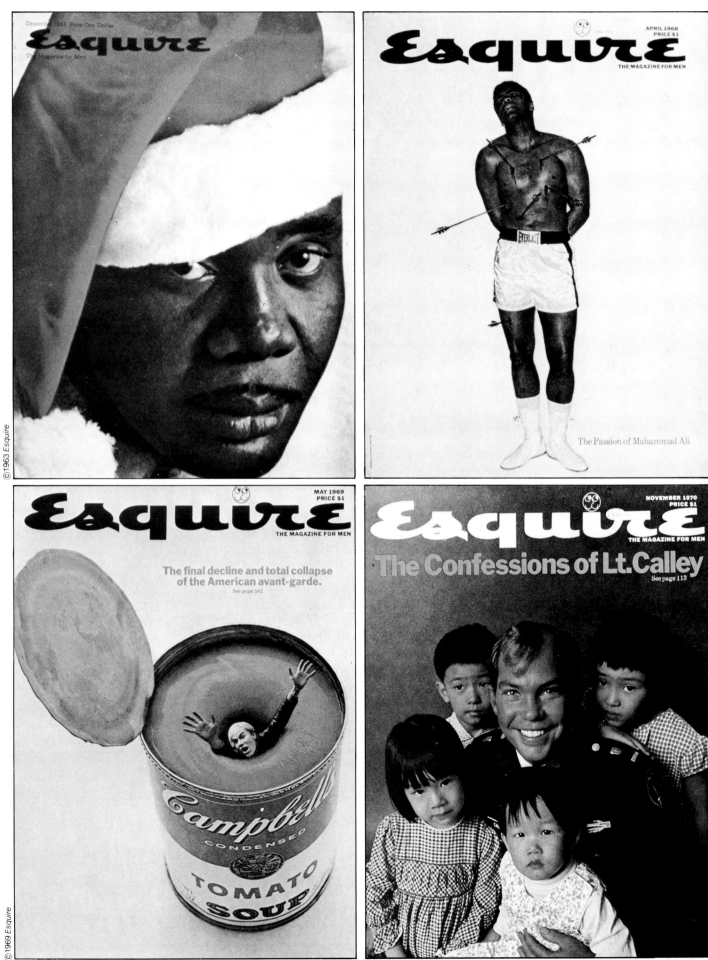

©1963 Esquire

©1968 Esquire

©1969 Esquire

©1970 Esquire

This new publication needed a cover for a lead story about the increasing use of nudes in advertising. Lois tried to get top executives from other agencies to pose for this creative conference, but finally had to fill the table with his own staff (Lois himself can be seen at the far end of the table). The two women are models, however; "Even I don't have the nerve to ask my female employees to pose bare," he said.

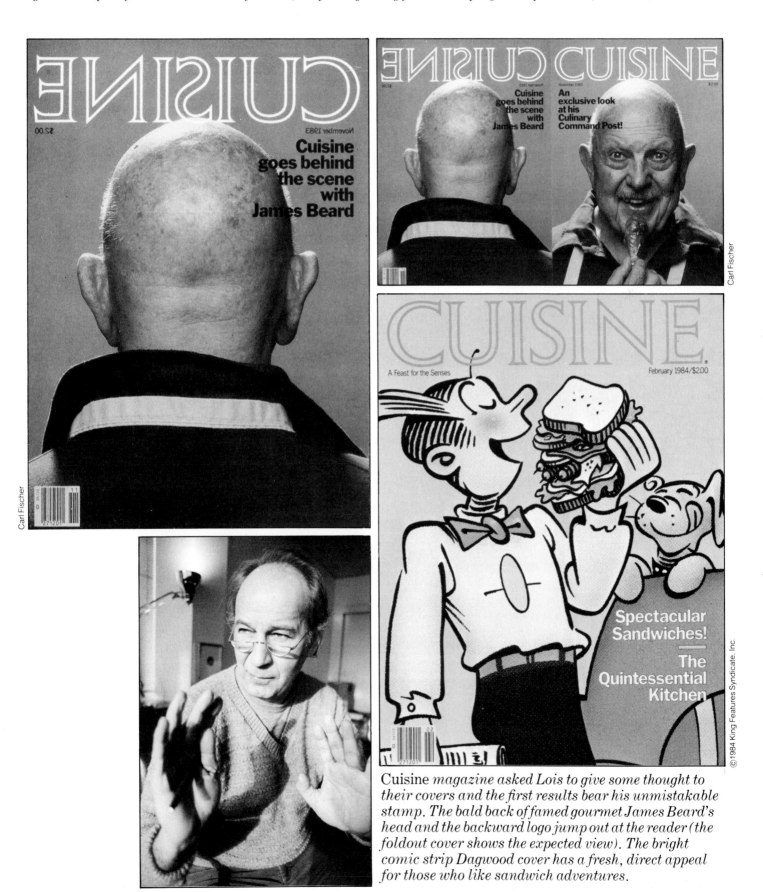

Cuisine *magazine asked Lois to give some thought to their covers and the first results bear his unmistakable stamp. The bald back of famed gourmet James Beard's head and the backward logo jump out at the reader (the foldout cover shows the expected view). The bright comic strip Dagwood cover has a fresh, direct appeal for those who like sandwich adventures.*

STAN MACK

Acute reporter and master of the minidrama, Mack literally roams the streets of New York in search of material for his regular columns in the Village Voice *newspaper and* Adweek *magazine. He finds abundant dialogue for his episodes in the restaurants, galleries, and offices of the city.*

A telling statement on human nature is this drawing of a little man in a knight suit. It can be read many ways: Is this a Walter Mitty dreaming of what he might be? Does it indicate that all of us could make more of a splash in the world if we had a little more courage? The drawing was originally made to advertise a bank. At right are three of the Adweek *adventures in the world of communication.*

GUY BILLOUT

Much has been written about Gallic wit, but until one sees a selection of work by this artist the meaning of the term is vague. Irony and sympathy for fellow human beings pervade Billout's drawings and paintings. The viewer is tempted to say to himself "Damn! If Voltaire had been able to draw, it would have looked like this." Billout claims that getting visual ideas is hard for him ("rather like giving birth"), and he dreads starting the process each time. The results suggest it is so difficult perhaps because Billout demands so much of himself; the idea must be unique.

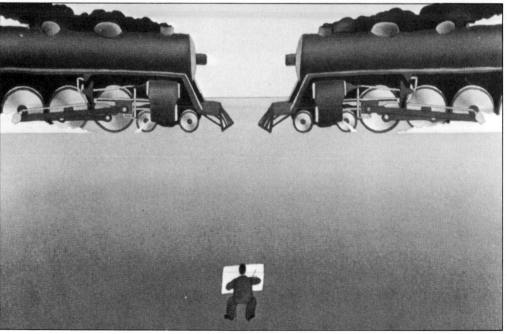

This self-promotion card places the artist exactly where he sees himself, as a detached chronicler of human drama.

Condemned to Death

The capital punishment lottery, by David Bruck

The New Republic *wanted to feature a story on the unjustness of capital punishment. The solution, a judge's gavel spraying blood to both sides, occurred to the artist while he was taking a break from the problem, walking in his neighborhood.*

A top drawer French hairstyling salon was opening a branch in New York. It seemed appropriate to make further use of another French gift, the Statue of Liberty, to make the announcement.

BILLOUT

Eyecatching incongruity makes these book illustrations stick in the mind. On the opposite page Noah's liner sails serenely past the spire of New York's Chrysler Building. At left, the climber who finally makes it to the peak of a mountain is dismayed to find it occupied — by a spider on a strand of web.

Billout is the temperamental opposite of George Lois. His work, which has some of the formality of a Hiroshige print, is created in a cluttered surrounding. Lois is voluble and demonstrative while Billout is quiet and reflective, but both share a fine sensitivity to people and the ways human beings react.

BILLOUT

Each table of contents page for the magazine Avenue *carries one of these drawings; each is a different observation on the theme of New York's Fifth Avenue.*

Mankind and the rest of the animal kingdom must find ways to live together. This is the admonition of the artist and a favorite illustration theme.

MILTON GLASER

There seems to be no limit to his excellence—illustration, print design, logo design—anything visual is fair game for Glaser. Recently he masterminded a top-to-bottom image overhaul for Grand Union, a chain of supermarkets; the program penetrated down as far as making custom labels for condiments. Glaser's most visible work, however, has been with posters, two of which are shown on this spread. He works from nine to six-thirty most days in the small brownstone townhouse that originally sheltered New York *magazine, the linear ancestor of all present city publications. Glaser says this studio is used "mostly for business—talking to clients, supervising projects with assistants, arranging jobs," but the more highly personal work, the drawings, are usually done either in his westside Manhattan apartment or at his country house in Woodstock, New York.*

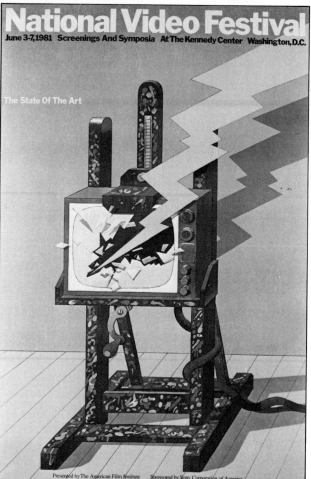

This poster was created for a national video technology fair in Washington, D.C. The visual image echoes the festival slogan, "The State of the Art."

Produced for a special entertainment event sponsored by the New York Zoological Society, this poster gets its beauty and uniqueness through the use of wild and arbitrary color on the snow leopard—teal blue, violet, tomato red, and mustard gold.

GLASER

Evolution of an image

The American Institute of Graphic Arts holds an annual show of the best print design done during the year, and Glaser was asked to design a poster for this event. Since the show would display items meant to be read, he decided that the principal image should be a person reading. Part of the solution was to impose a severe light and dark modeling on the subject, a chiaroscuro effect with no middle tones. The sequence of sketches on these pages shows the shifts in viewpoint and changes of composition used to reach the final form. The finished poster was printed in rust, spruce green, and black on white paper.

5

6

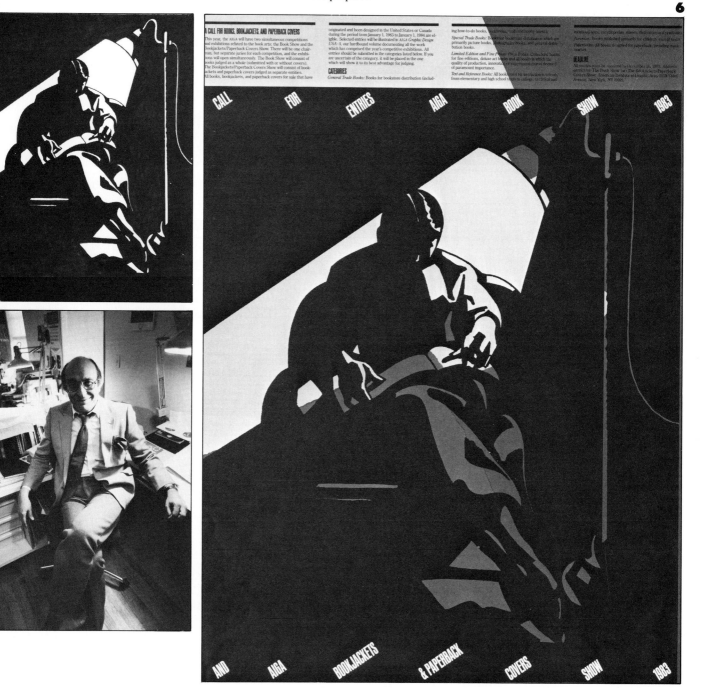

GLASER

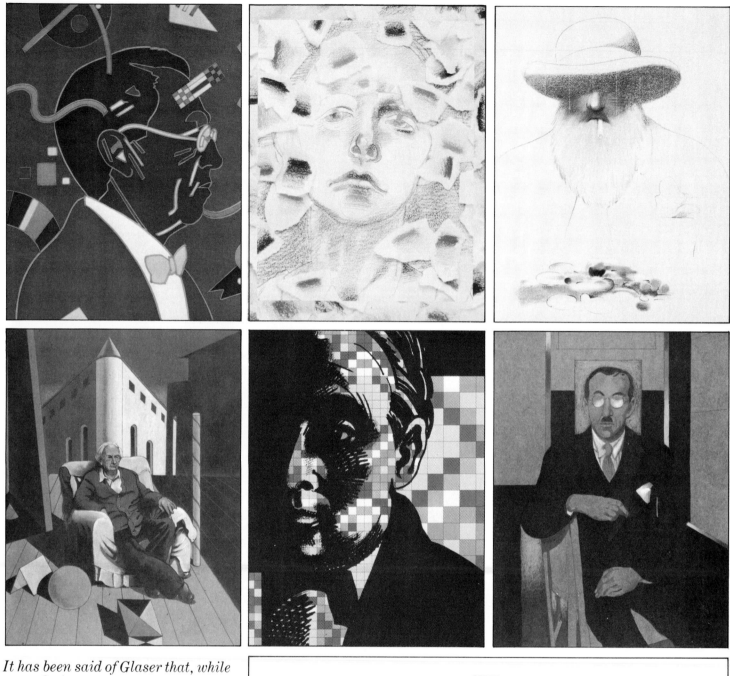

It has been said of Glaser that, while some designers seem to wear many hats, he has the additional ability to change hands—to put on a different pair of hands when a new drawing style is needed. This calendar offers a fine demonstration of his technical virtuosity; each image pays homage to one of his favorite artists, whose painting or drawing style he emulates.

KEY					
Wassily Kandinsky	Georgia O'Keeffe	Claude Monet	Kitagawa Utamaro	Edvard Munch	Sonia Delaunay
Giorgio DeChirico	Paul Klee	Piet Mondrian	Odilon Redon	Max Ernst	Gustav Klimt

SEYMOUR CHWAST

*A key figure in modern illustration,
Chwast has several imitators, but only his
work shows the magical blend of childlike
directness and sophistication from piece to
piece. Chwast is a quiet and modest man,
not at all the type of person the exuberance
of his work would lead one to expect. He is
the only founder of the original Push Pin
Studio to remain with the firm, but the
creative shop and the talent that made it
great are as strong as ever.*

*A good sample of the
appealing whimsy of a
Chwast design is this
poster made for the
Fort Lauderdale Art
Institute, a vocational
school for commercial
artists and designers.*

An insert cover for a special section of Adweek *devoted to creativity shows an ad maker aboard a fantastic cycle, going headlong to an uncertain destination.*

Chwast's special creative touch is shown in this 1982 peace march poster. The dove becomes an umbrella of common purpose for people from all walks of life.

CHWAST

This poster, created for the famous BBC television series about power struggles in imperial Rome, is a classic piece of graphic design.

The cover of a guide to New York writers' markets makes a building into a fountain pen.

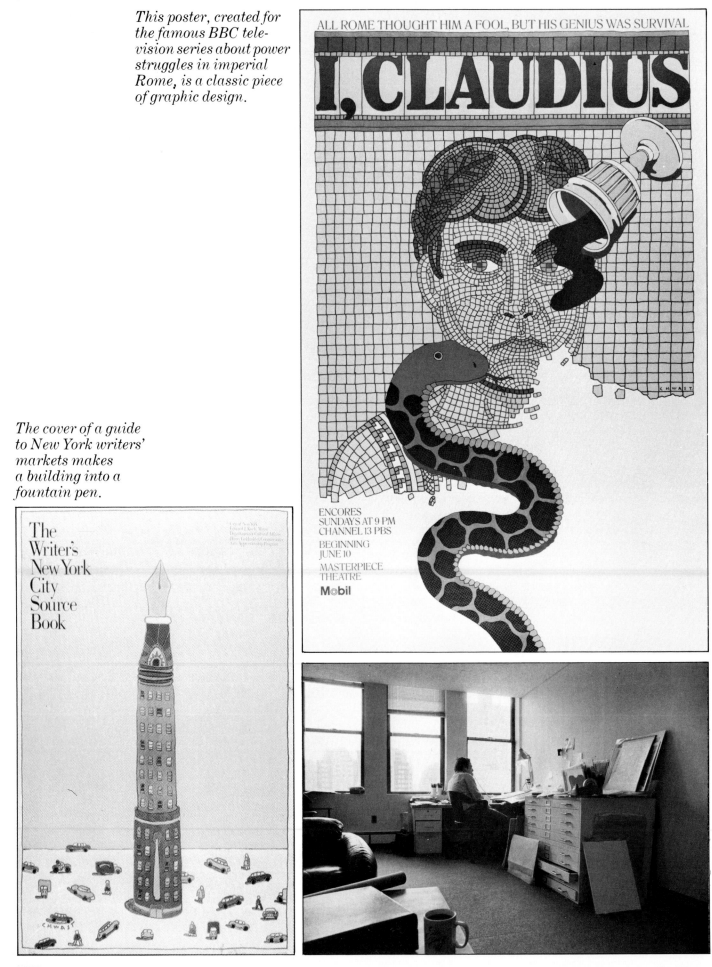

Shown are three of the illustrations promoting Forbes *magazine, "the Capitalist Tool." Ads for the financial journal were designed for display in magazines and as large cards for commuter railroad billboards.*

PAULA SCHER

For many years, Scher was an art director at CBS Records. Some of her many album cover designs, shown on these pages, testify to the instant clarity and quality of her skill. She says she works best when there is another good designer in the same room with whom she can interact; this need was fulfilled when she went into partnership with two other talented art professionals, Richard Mantel and Terry Koppel. They share a large room in a building set among warehouses in the middle of Manhattan. Gregarious and articulate, Scher is rated highly by the illustrators and photographers with whom she works; they appreciate her willingness to allow freedom for them to contribute to the concepts. She invites them to think.

Richard Hess

Robert Giusti

Band instruments are caught among trucks and taxis. This painting invites careful exploration, but the idea is quickly grasped.

Two gentlemen stroll in the park, meet, shake hands. The image demonstrates a true meeting of minds.

Arnold Rosenberg

An actual manhole cover was photographed and a color print was retouched to form a heart shape.

David Wilcox

The scene resembles the prison exercise yard suggested by the title of the album.

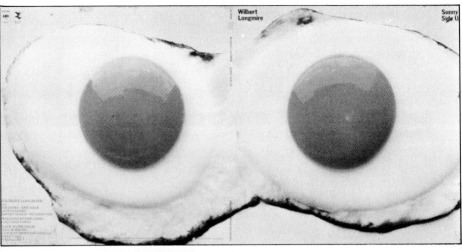

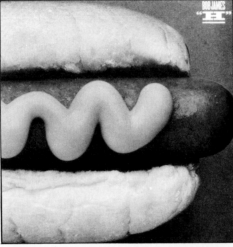

The client, Tappan Zee Records, asked for a family look for its series of jazz albums. This was achieved with super still life photographs that played on the album names.

ELWOOD SMITH

Few artists today work in a more appealing style than Elwood Smith. His illustrations are in heavy demand for two major reasons. First, there is the old-shoe comfort of a familiar style of drawing. It is strongly reminiscent of that of Smith's hero, George Herriman, creator of Krazy Kat, and its pull owes much to simple nostalgia. It evokes simpler, more human times. Second, the conceptual vigor he gives the individual pieces makes them unique. They are far more than slavish copies of an earlier technique; the thinking is original and entertaining. The artist is much like his work: open, outgoing, uninhibited, and unpretentious.

Affection for his pets shows in these two assignments. A cover for the Chicago Tribune *magazine demonstrates the proper treatment for man's best friend, while the cat-draped reader was featured in an information sheet for pet owners.*

SMITH

Smith produced two sketches for
Datamation's *report on the proliferation of*
computer software companies, but the
magazine chose to show individual rather
than grouped lemmings. At right is a study
sketch of a man on a runaway luge; the artist
identifies strongly with this figure.

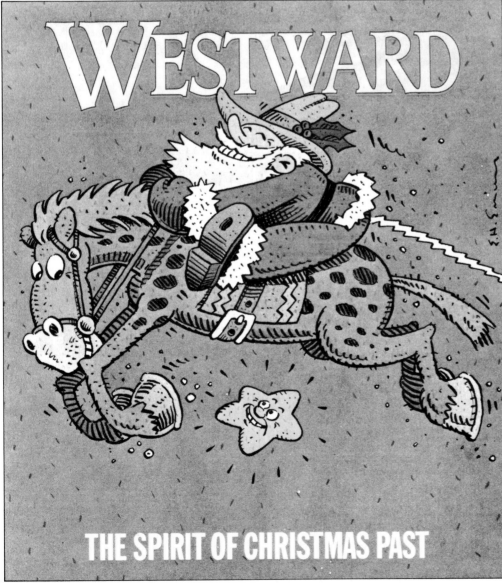

Two different comments on the Christmas season are made by these covers for the Sunday newspaper supplements Westward *and* Chicago Tribune Magazine.

SMITH

ASPHALT JUNGLE FEVER... *Manhattan apes the tropics. With warm temperatures and a rising sea level due early next century, will the urban jungle become the real thing? Artist Elwood H. Smith predicts a primate climate that'll put high tech on ice. And we'll have a hot time in the old town...*

Vanity Fair *speculates on changes resulting from warmer weather that will be caused by the melting of the polar ice caps; Smith gives form to the new urban jungle.*

Great hot dogs in the Chicago area are the theme of this cover.

What happens if all
adhesives suddenly give
way? The world falls
apart, of course.

SMITH

Just where is Elwood H. Smith, anyhow?

Smith's awareness of news trends and life-style changes is demonstrated in these two works. At right is a comment on what can happen to a slogan on its progress from creative team to client in an advertising agency. Smith recently moved from New York City to idyllic Rhinebeck, a pretty little German-settled village up the Hudson River. The mailer booklet below was created to reestablish contact with his clients and friends.

REACH OUT AND TOUCH SOMEONE!

He's not gone underground.

He's not crossed over to the other side.

He's not gone to the Falklands.

He's not gone fishin'.

He's not become a recluse.

Elwood H. Smith
has moved to:
2 Locust Grove Road
Rhinebeck, New York
12572

ANTHONY ANGOTTI

Top advertising design is increasingly dominated by designers who deal with equal deftness in verbal and visual ideas. Angotti is a good example of this type of rounded professional communicator and his creations show that he is an accomplished student of human nature.

This print series for BMW takes advantage of anxiety about the latest economic trends to make its powerful pitch.

LAST YEAR, A CAR OUT-PERFORMED 318 STOCKS ON THE NEW YORK STOCK EXCHANGE.

BMW UPDATES ITS ANNUAL REPORT TO PERFORMANCE-MINDED INVESTORS.

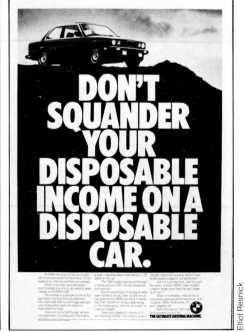

DON'T SQUANDER YOUR DISPOSABLE INCOME ON A DISPOSABLE CAR.

THE ULTIMATE DRIVING MACHINE.

Michael Sarasin

ANNCR: (VO) At Club Med you can water-ski...

...play tennis...

...snorkel...

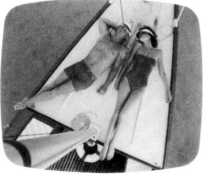

...or sail.

Windsurf...

...or play volleyball.

At Club Med you can exercise everything.

Including your right not to exercise anything.

CLUB MED
The antidote for civilization.

SONG: The Club Med vacation. The antidote for civilization.

Among the most memorable television spots is this offbeat ad for Club Med resorts. The expected approach might have shown hordes of attractive young people endlessly playing sports of their choice...

Judy Carter

Judy Carter

Craig Perman

NANCY RICE AND TOM McELLIGOTT

Rice, as executive art director, and McElligott, as head creative director, have combined their talents to create some of the strongest statements since George Lois' Esquire covers. Their work's power relies heavily on verbal color and offers the best proof of the impact of "street language," headlines that reflect the ways in which people actually talk. The messages are unmistakable — and unforgettable.

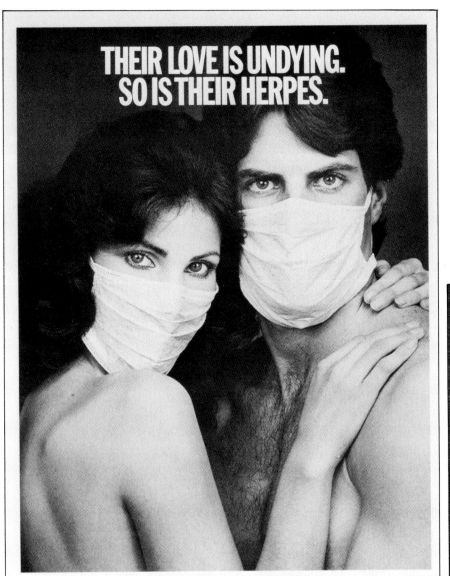

THEIR LOVE IS UNDYING.
SO IS THEIR HERPES.

For about 20 million Americans, genital herpes is here to stay.
As diseases go, it may not be the most serious, but it can have devastating psychological effects
on those who catch it — partly because there is no known cure for it, and partly because
of the shame, guilt, social stigma and ancient taboos associated with venereal diseases. The herpes epidemic
is currently making 350,000 Minnesotans more or less miserable. Tonight, watch TV 11's
exclusive one-hour report, "Herpes is Forever," hosted by John Bachman.
Parental guidance is recommended because portions of the program are graphically explicit.

WATCH "HERPES IS FOREVER" TONIGHT, 9-10PM.

NewsCenter 11
THE ONES TO TURN TO FOR NEWS.

Craig Perman

Two newspaper ads done for a local television station hammer home points about two special broadcast reports: one on drug abuse (left) and one on the latest social disease (above).

Judy Carter

Posters created for the Minnesota Nuclear Freeze Campaign dramatize the effects of atomic warfare.

IF THIS IS WHAT
HAPPENS TO A DOLL
10 MILES FROM
A NUCLEAR BLAST

GUESS WHAT HAPPENS
TO ITS OWNER.

MINNESOTA NUCLEAR FREEZE CAMPAIGN

Kerry Peterson (Marvy)

SURE THE WORLD
CAN SURVIVE
NUCLEAR WAR.

IT'LL JUST LOOK
A LITTLE DIFFERENT.

MINNESOTA NUCLEAR FREEZE CAMPAIGN

RICE AND McELLIGOTT

A local Episcopal church group came to the agency asking for ads and posters that would attract new members. The strategy that developed was to position the Episcopal Church as a moderate, thoughtful, and reasonable worship alternative to fundamentalist and television Christianity. Says Rice, "It was fun because the clear positioning and strong concepts were never bitten to death by the usual array of committee guppies. Since the budget was small we were able to do what we wanted."

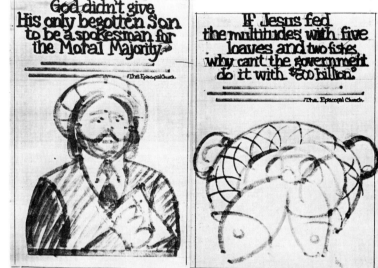

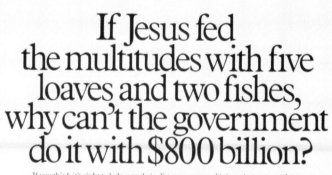

God didn't give His only begotten Son to be a spokesman for the moral majority.

If you think Jesus loves all people — even those who don't agree with Him — come and join us in a service where diversity is not only allowed, but welcomed.
The Episcopal Church

If Jesus fed the multitudes with five loaves and two fishes, why can't the government do it with $800 billion?

If you think it's right to help people in distress, come and join us in an atmosphere where compassion toward people and the worship of God come together in joy and fellowship.
The Episcopal Church

Tom Bach (Marvy)

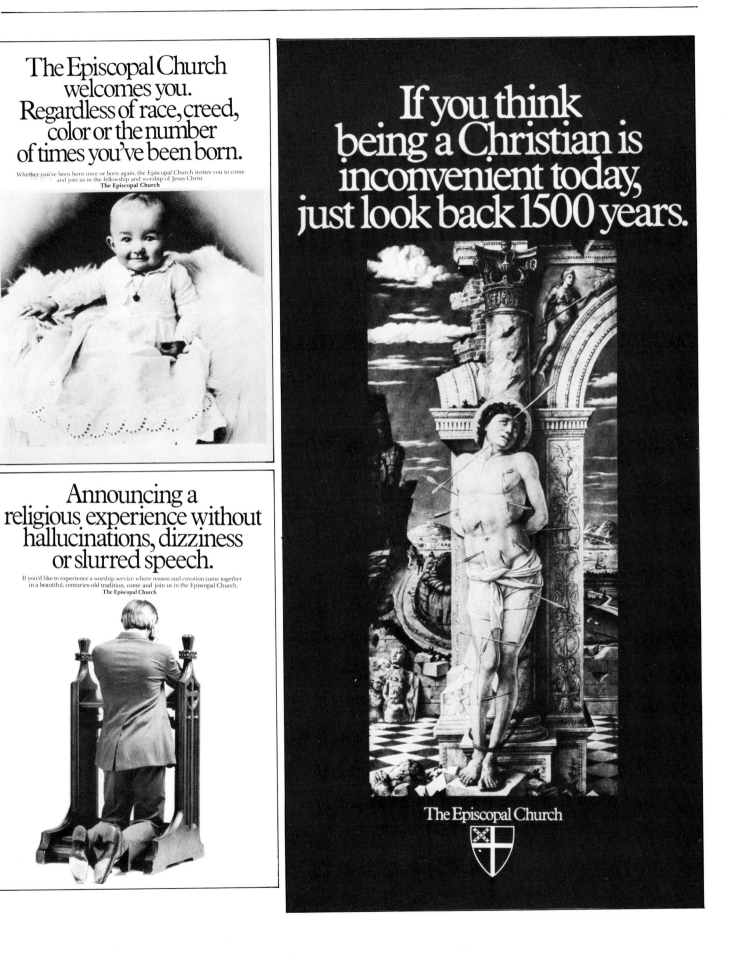

The Episcopal Church welcomes you. Regardless of race, creed, color or the number of times you've been born.

Whether you've been born once or born again, the Episcopal Church invites you to come and join us in the fellowship and worship of Jesus Christ.
The Episcopal Church

Announcing a religious experience without hallucinations, dizziness or slurred speech.

If you'd like to experience a worship service where reason and emotion come together in a beautiful, centuries-old tradition, come and join us in the Episcopal Church.
The Episcopal Church

Jim Arndt/Berthiaume

If you think being a Christian is inconvenient today, just look back 1500 years.

The Episcopal Church

RICE AND McELLIGOTT

Whose birthday is it, anyway?

Considering the fact that Jesus had his doubts, why can't you?

The Episcopal Church

One-on-one personal appeal is exemplified by these posters, two of the most recent ones done for the Episcopal Church.

Frank Miller

This poster was designed for a chain of hardware stores in the Minneapolis area. Note the venerable figure of speech, used with a twist of meaning.

These two ads for ITT Life Insurance make their points quickly and irresistibly.

Scott Baker

Dick Jones

BRUCE BACON

A designer, photographer, and painter, Bacon has put most of his energy into gallery work in recent years. He finds the freewheeling association highly rewarding—his conceptual paintings are being snapped up by private collections around the world. An avid student of creativity and an inspiring design teacher, he has been characterized as a cross between Leonardo and Rube Goldberg. Some of his sky motif paintings are shown on this spread.

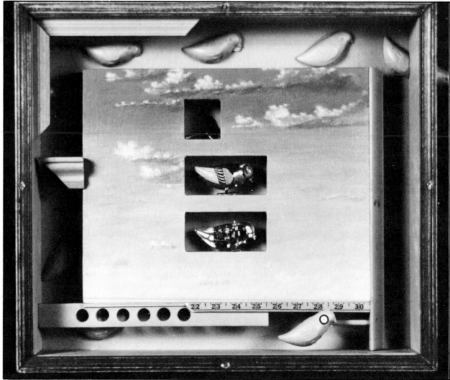

Bird Game in a Box.

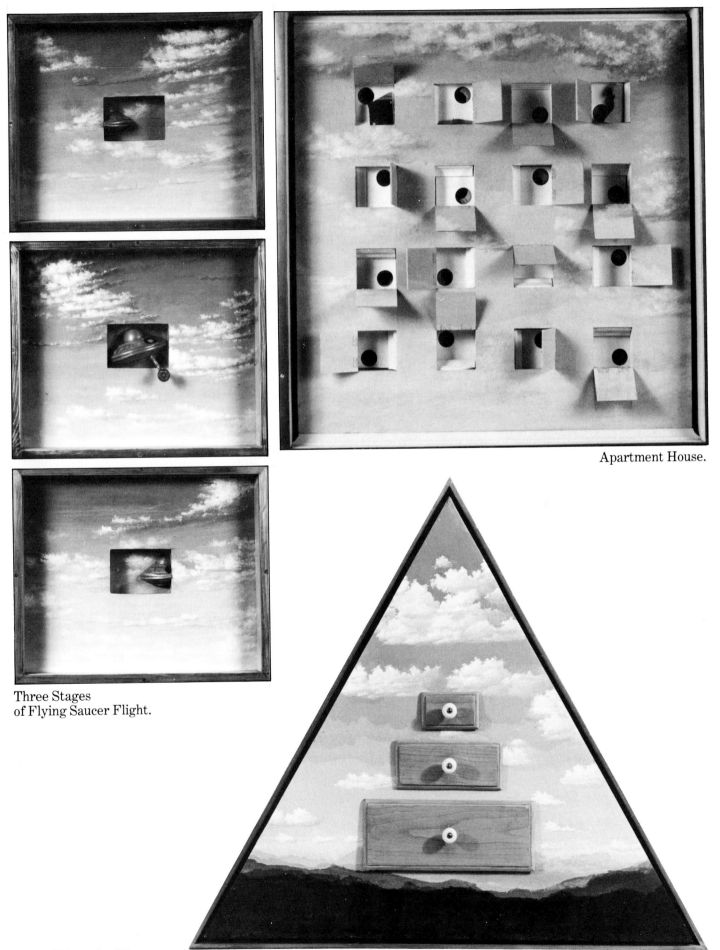

Apartment House.

Three Stages
of Flying Saucer Flight.

Triumph of Reason.

BACON

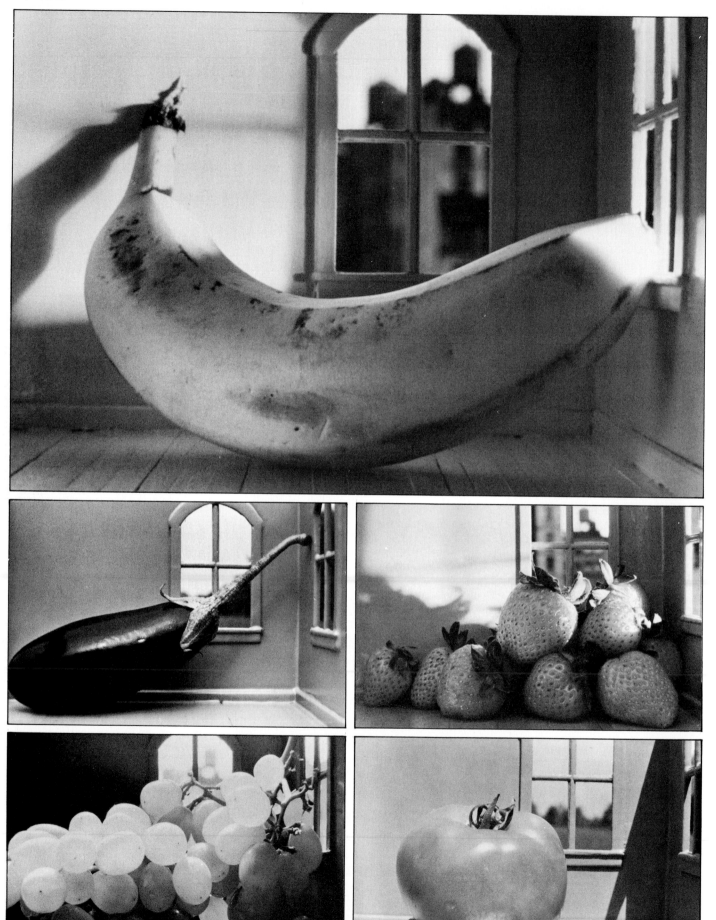

*At left is a series of
experimental photographs
entitled* Fruit of the Room.

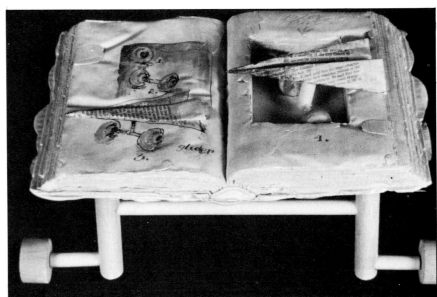

*Three of Bacon's pieces of
sculpture are shown here.
At right is* The Glider Book
of Spare Parts, *below is*
The Coffeetable Book, *and
at bottom is* The Sky-
writing Book.

DIK BROWNE

Creator of the comic strip Hagar the Horrible, *Browne works at a pace that produces twenty to thirty ideas each week. The central character in the strip is a lovable, walruslike Viking who commutes by longboat each day to his job . . . raiding and pillaging. The situations generally mirror areas of minor frustration common to most people and, as Browne readily admits, the humor has its roots in tragedy—the basic paradox of human existence, the seeming impossibility of complete happiness. In the photo above, Browne is shown working in his basement studio with laundry flapping around his head. At right he shares a melancholy coffee break with a friend.*

BROWNE

The syndicated strip is something of a cottage industry; a portion of the drawing is done by Dik's talented sons, Chris (left) and Chance.

Browne describes
feelings most of us have
about certain professions
with surgical clarity. On
these pages, lawyers and
salesmen are brought
under scrutiny.

CARVETH HILTON KRAMER

Originally trained as an illustrator, Kramer drifted into staff design for publications and found that the satisfactions of editorial work suited him well. His ability has brought him a large collection of awards from national art and design competitions. Years of generating ideas for the pages of Psychology Today have left Kramer with an ingrained intellectual restlessness, a need to project, to plan, and make new combinations.

This is a sketch for an article about why people fear dentists.

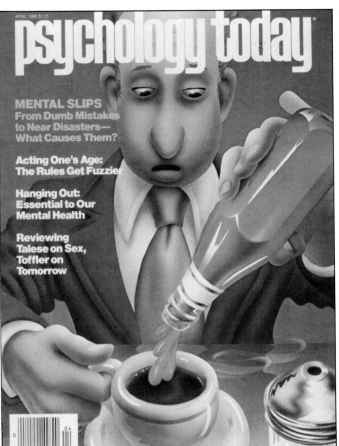

Robert Grossman

At left is Kramer's solution for a cover story on dumb or careless mistakes. Below are the planning sketch and finished opener for an article on the widespread fear of skiing.

FEAR OF SKIING

On the slopes—or in any high-risk situation—fear can work for you if you know how to make constructive use of it. Of course, according to the masters of the "inner game," you first have to understand what fear is all about.

By W. Timothy Gallwey and Robert Kriegel

From the book Inner Skiing by W. Timothy Gallwey and Robert Kriegel, published by Random House. Copyright ©1977 by W. Timothy Gallwey.

78 PSYCHOLOGY TODAY NOVEMBER 1977

Seymour Chwast

THERE ARE TWO KINDS of challenge in skiing: the external challenge of the slope, and the doubt and fears within the skier. Those who think of skiing exclusively as a matter of perfecting technique and proper form miss at least half of what the sport really is. Skiing is more than a parallel turn; it is an opportunity to go beyond some of the internal limitations we place on our ability to learn and enjoy anything we do.

Fritz Perls, the father of Gestalt therapy, observed that human beings are members of the only species that possesses the capability to interfere with its own growth. As a result, most of us are performing far below our potential. When attempting to improve a skill like skiing, we practice various patterns of interference. Voices tell us that we can't ski well, so we don't; that we'll probably fall on the next mogul, so we do; that we should remember at least a dozen dos and don'ts to insure good form, which is totally impossible to do while skiing.

What part of ourselves does this interfering? We call it Self 1. You can hear Self 1 chattering within your head, criticizing, judging, worrying, and pontificating about things it doesn't really understand. It distorts perceptions and blocks the expression of the natural potential of the body, which we call Self 2. The point of inner skiing is to decrease the interference of Self 1 and to allow Self 2 to express itself fully.

At a recent inner-skiing workshop, we asked 300 participants to name the internal obstacles that most interfered with their proficiency and enjoyment of the sport. By far the most frequently cited problem was fear. Nearly everyone considered fear at least detrimental to their skiing, and most saw it as a major problem. Some even said that because of it, they had never wanted to learn how to ski. Most were hoping for a magical cure—something that would create instant courage.

The first question to ask, however, is whether all fear is bad. Is there a kind of fear that isn't harmful but, in fact, helpful? Haven't most of us been in a real

Above is Kramer's planning sketch for the cover at right, an issue on dream research conducted by behavioral scientists. On the opposite page is the photoillustration for an article on keeping secrets.

Marvin Mattelson

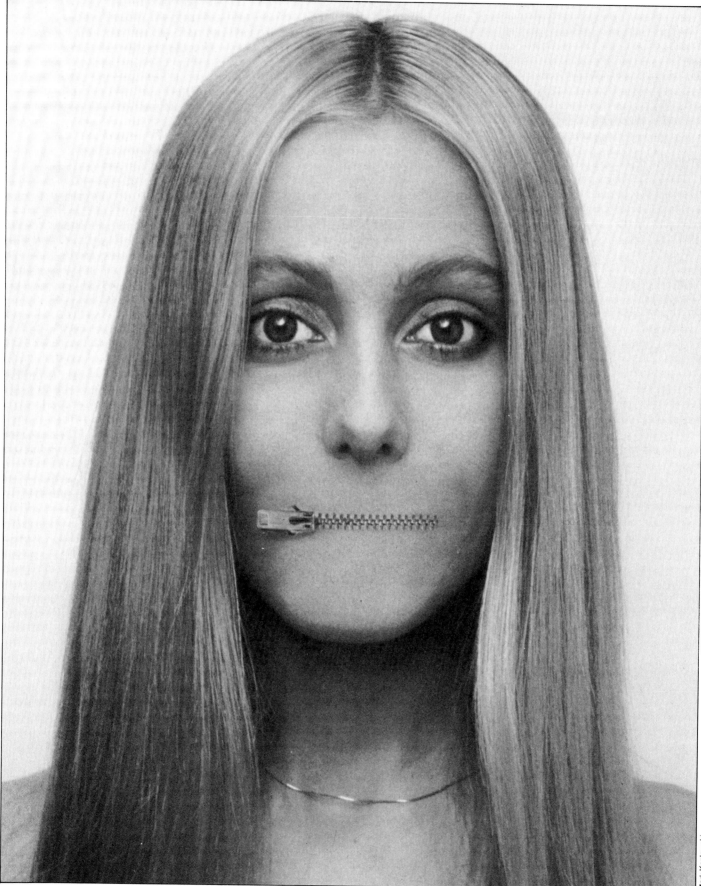

Mark Kozlowski

Use humor freely. All of the panelists acknowledged their debt to humor as a tool of creativity, and all considered it enormously important: "Central to my work," said Smith; "The first thing I reach for," said Billout; McElligott: "I love irony and most other types of humor. Laughter is a pretty reliable way to win people over." And Nancy Rice adds, "Humor disarms. Almost everyone responds to portrayals of human foibles." Paula Scher: "I'm always using puns. Humor is my first resort; I want most of my work to be witty or even funny."

Most creative people I've ever known have been incorrigible jokers, full of whimsy and ever ready to laugh, even at themselves—especially at themselves. Humor has great direct benefits to creative work; it can make an impression stick. True, humor is not appropriate in many solutions, but different forms of it abound in today's communication. Try to remember some television spots you've seen, and chances are you'll find yourself reliving the Alka Seltzer commercial in which endless retakes show an actor who's supposed to deliver the line, "Mama mia! That's some spicy meatball!" eventually reduced to babbling "Meesy . . . micey . . ." Or perhaps you'll think of the monks rediscovering the "miracle" of the high-speed office copier and gazing with gratitude toward Heaven. Or perhaps you'll picture the Federal Express commercials directed by the Norman Rockwell of television directing, Joe Sedelmaier, the spots in which everything in an office is said and done at machine-gun speed ("In this fast-moving world of ours . . ."). And who can forget the tough little old lady demanding "Where's the beef?" Humor is the all-purpose lubricant of our society, and messages delivered with its grace will generally outlive the others.

The humor must not be hurtful, of course. Use of irony (as long as it's not directed at the viewer), belly laughs (as long as you're not laughing at the reader), or whimsy (as long as your audience finds it entertaining) are wonderful tools, but the humor must be tempered with compassion for your fellow human beings. Jibes at human weakness must be delivered with a sense of kinship. The wit must not be simply mean.

How to cope with blocks. Most of the panel members said they didn't try to force their way through blocks. When a problem defied solution, when none of the ideas seemed fresh enough, most of the designers walked away from the problem for a while. Most of them changed the scene—went for walks or switched to some other activity. (The one exception was George Lois. Whenever solutions fail to appear, he goes back to pick at the original research, to find the additional small piece of information that will spark an idea.) Glaser says one of his favorite escapes is to the highly detailed and colorful pages of a book on eighteenth-century Indian miniature paintings. Other panelists had their own recipes: Smith peruses the Krazy Kat strips of his childhood idol, George Herriman, to "see how he broke up the space, how he timed the action." Billout says he finds it effective to turn to some mundane chore—washing dishes or dusting furniture—something that will show tangible results when done. McElligott goes jogging or watches a taped movie at home. Mack calls people he knows to open other lines of investigation. It's enough for me to get up and go for a cup of coffee or find someone with whom to chat for five or ten minutes; when I return to the problem I can usually resume work. The consensus: leave it alone for a while, let the problem sit for five minutes, an hour, a day or two days. This is almost sure to yield more usable material than trying to grind out the idea at one sitting.

Keep high standards for your ideas. When you find yourself looking at final doodles that represent the best of your visual solutions, you should be asking questions that an interested outsider might ask. Is it different? Does it dramatize the situation or product enough? Some of the panelists had their own requirements. McElligott asks himself and his team, "Have we gotten as near the edge as we could? Have we been as daring as we could without slipping into the range of bad taste? Or have we been too safe?" George Lois always measures a solution by finding out whether or not it scares the client. If it does, he has added faith in the idea. (Ask the man who has shown Sonny Liston as Santa Claus, Muhammad Ali as St. Sebastian, and a smiling Lieutenant Calley of My Lai fame in a group portrait with Oriental children.)

Kramer asks himself, once he's dismissed his internal task force, whether the resulting idea surprises him. Mack, of course, asks whether the idea he chooses stings a little. Is it poignant? Dik Browne submits his best choices to members of his family and responds to a rating system with which his wife and sons mark each rough: 1 means "use immediately," 2 means "needs help," and 3 means "lose it." (He denies rumors that there are lower grades such as "go to your room" or "leave town.")

Even if you feel you're the best judge of the quality of your idea, you must dissociate yourself from the problem. You must see it as if someone had just shown it to you for the first time, and you must ask yourself whether the message has impact under these circumstances. This detachment is not easy. Since your idea must eventually appeal to ordinary people—to readers or viewers—it is probably wise to do a limited preview of the best ideas with others who haven't been involved in the creative process. These first reactions can tell you whether an idea conveys the message you intended and can indicate how quickly—or slowly—it does so. And remember that some of those who see your favorite solutions may not like one or all of them. When this happens, keep your cool; instead of reacting defensively, try to find out the basis for any adverse reaction, ask yourself if there are shortcomings of meaning or implication that may be causing trouble for your critic. Some calm, undefensive questions should reveal whether an idea really needs adjustment or whether the negative reaction is a result of a personal bias on the part of the viewer. Whether the responses are favorable or not, some will ring true; some may echo your own half-submerged doubts or, on the other hand, bolster your own opinions. Take the comments for what they are—just another factor in your choice of a good final verbal and visual solution.

These are the immediate considerations that become important when you get down to work on a specific problem. But, more generally, are there ways to become a better graphic designer? What goals and conditioning would we prescribe for a young artist who wants to develop the skills needed to succeed in the world of professional art? What sort of program would I or my panel of professionals recommend to someone starting a career in communications? By consensus, these are some of the general pointers we have to offer:

Become a student of human nature. The good communicator must become a lay psychologist. You should learn what motivates people and what they want, and then play on this knowledge to make your readers respond. You build this understanding by observing the world around you and by watching people at work and at play.

Be an intellectual gypsy. Explore as many different areas of human endeavor as you can; find out all you can about what other people do. Paula Scher expresses this concern for the students in her design classes: "I worry that, bright and imaginative as these kids are, their fund of personal information is too low. Conceptual capability only comes after a certain amount of browsing in human affairs; how can a designer make a significant statement unless he or she has these reserves of information? I look at everything. I read a lot and habitually poke around the big bookstores in the city. And I encourage my students to do the same."

Professional communication offers a wonderful opportunity: the chance to explore lots of subjects. My own design experience, for example, introduced me to experts in the natural sciences who were authors of books for Time-Life, while other book projects got me involved with gardening, cooking, and travel. Art-directing *True* magazine enabled me to work with top writers; then the *New York Times* showed me the world of big money sport, golf, and tennis; and now, at *Medical Economics*, I find myself neck deep in the doings of doctors and surgeons. All are fascinating worlds, and all have contributed to my breadth as a communicator. Ben Shahn, the painter, used to advise his students to do everything—everything that isn't illegal or harmful. His advice is still sound, because what we make comes from what we have: the more knowledge of the world we store, the richer the matrix upon which we can draw to make new combinations in design.

Ken Robbins, chairman and chief executive of one of the largest advertising agencies, urges his creative people to stretch their lives every available way. He says, "Nothing is more essential than an active curiosity. The creative process is nothing more than a function of all your observation and knowledge, constantly resynthesized and reapplied in terms of your own personal experience. Therefore, one can never read enough books, watch enough movies and stage plays, watch enough television, understand enough political positions, etc., in the search for fresh and surprising ideas."

Be a cultural catfish. The best communicators are those who study the unwritten history of society, the pop culture of the day and of days gone by. They collect bits of trivia and gobble up the castoff facts of society for later retrieval as needed to make the magic of fresh verbal and visual statements. They collect oddities and squirrel them away for possible future use.

Develop an ear for "street language." The term as coined by George Lois refers to the way people really talk to each other. The aspiring communicator must be able to echo the terms and rhythms of language as it is spoken in homes and offices and on the street. Kramer, in fact, says his favorite place for language research is the office building or department store elevator. He claims that being in an elevator seems to release inhibitions, that people on elevators tend to talk freely, and, since there is no hope of not being overheard, they seem to ignore eavesdroppers.

The skilled art director or writer knows he or she can reach more people faster by using familiar terms and imagery, the chatter of everyday life. Television is a good laboratory for developing this sense of language. I would go so far as to prescribe watching a few of the serials that catch the sound of our society. Listen to reruns of "M*A*S*H" and "Lou Grant," see how Archie Bunker and his family use and misuse words, catch some episodes of "Hill Street Blues," "Dallas," and "60 Minutes"—listen to what the characters say to each other and how they say it.

Street language is also to be found in newspapers, particularly in small-town or local journals. Depending on where you live, read the *Stamford Advocate* or the *Waterville Telegraph*.

The quest for street language should include some reading of syndicated comic strips. There's heavy value in the speech balloons of *Hagar the Horrible*, *Peanuts*, and *Garfield*. (Garfield says, "Show me a good mouser and I'll show you a cat with bad breath." Or "Life is like a warm bath. It's nice while you're in it, but the longer you stay the more wrinkled you get." Is there anyone who doesn't respond to language like that? I doubt it.)

Develop your sense of humor. There are solid reasons why creative people are frequently humorists. The ingredients of humor—illogic, perversity, odd combinations, contrast, irreverence—are devices used to surprise the audience. They are used to startle the hearer with an unexpected turn of cause and effect. The designer uses the same types of devices to stimulate an audience. You needn't become a Groucho Marx to be a good designer, but the habit of seeing humor in most topics seems to condition the innovative artist to seek unusual, memorable ways of saying things. A humorous bent seems to lead to valuable mix-and-match thought patterns.

Think of yourself as a showman. This kind of work—producing magazines, making television spots, creating pieces of promotion—has much in common with show business. This thought should not offend an editor or art director; very simply put, editorial showmanship is the craft of making readers respond as you wish them to. You carefully guide their attention from one thing to another, from one event to the next, until they see what you want them to see. They may even believe what you want them to believe after they see the climax of your performance. Sell is not a four-letter word to avoid; a moment's reflection will tell you that each of us sells almost every minute of his or her life. We sell our appearance to family and friends, we sell our ideas to our colleagues, and we sell ourselves to ourselves—that is, we do certain things to maintain our self-respect. I point this out for the benefit of anyone who chooses to enter editorial rather than advertising design because the former is "purer"—because it depends less on selling something. Not so.

Keep company with the best. Study the annual award selections that are published each year—the One Show yearbook, the Creativity annual, the New York Art Directors Show yearbook, the Society of Illustrators award collection, the Graphis Annuals and the Print Casebooks—roundups of the very best, most effective solutions, the best-angled ideas being done each year. Most of the panel of top professionals said they drew inspiration from these regularly. Tom McElligott said they set the standard for his own work and that he found it profitable to leaf through them once in a while as a way of building excitement before attacking a problem. Angotti stressed the importance of being with creative colleagues as represented by their work in these books. Some of the panelists used these books as diversion and inspiration when faced with blocks, but their main value was seen to be general and continuing; they are used as a constant source of stimulation.

Read all you can. Fiction or nonfiction, history or romance—almost any kind of literature can broaden your experience. You can use the authors' eyes to extend your own view of human nature and the world.

Attend a good art school. A few years of training in the best art school you can afford can give you many of the mechanics of the trade as well as a limited amount of critical exchange. However, this training is only preliminary, for real growth only starts after school—in the real world, with the real concerns of a real job.

Ongoing training: pushups for the brain. Finally, the associative and synthesizing skills a designer needs can be cultivated and sharpened with a couple of simple drills. Just as you would lift weights to strenghten your arms or run to improve your heart, so you must build your brain in special ways. The brain, like any other part of your body, responds to specific demand—becomes stronger the more you ask it to do—and, if you wish your brain to react more promptly when there is creative work to do, you must give it a kind of exercise tailored to enlarge this capacity. The exercise should have some of the same functional characteristics of the work for which the brain is needed; in this case, the exercise should consist of putting disparate things together, of making new and different things out of existing ones, of associating objects and symbols with different contexts.

There are many such drills that you can devise yourself. You could, for instance, play Tony Angotti's "what if" game. As you go through the day, pick objects around you—a quaint building, a desk lamp with a flex-style arm, a boulder—and imagine fantastic, outlandish uses of these objects. Lift each thing into an alien surrounding; make it do a job for which it surely wasn't designed; change its context.

Or you can pick a fellow passenger sitting across from you in the subway and build three short descriptions of different lives he might fit—three roles he might play—based on what you see: his clothes, what he's carrying with him, how he carries himself, or how he sits. Look for clues to the kind of character he might be. (Warning: don't look too long, especially in a subway train; here, as elsewhere in the animal world, a direct stare is a challenge.)

The exercise I have constructed need only be done a couple of times a year. First you choose a general subject about which you know a reasonable amount: office etiquette and politics, marriage, history, child psychology, a sport you play, etcetera. This becomes a matrix from which you will make imaginary articles, a frame of reference with which you will create opening spreads of magazine stories. Then, when you have chosen your subject area, leaf through any of the annual graphics books mentioned above—the Society of Illustrators annual, the Graphis Annual, a Photographis Annual—and let randomly selected images trigger ideas within your subject area. For example, if tennis is your subject and you see a photograph of a hen sitting on top of a clear wine bottle full of unbroken eggs, this may suggest an article about keeping tennis balls fresh and lively. The spread you sketch may show three wine bottles of slightly different shapes, each containing a single tennis ball. The headline you write and add to the sketch might be, "Is there *no* way to keep a tennis ball fresh?"

To demonstrate the idea more completely, the sketches resulting from two such drills appear on these pages. The subjects I chose were tennis and children. The sketches, though quite rough, are done at something close to actual page size—that is, 8½″ × 11″. They contain just enough detail to convey the idea. Normally, one of these art annuals should produce enough inspiration for anywhere from half a dozen to a dozen of these imaginary articles. This drill need only be done a couple of times a year, less if your job already entails a large amount of conceptual work.

Another useful conceptual exercise was devised by Bruce Bacon. This is a drill for visualizing addition and subtraction by depicting objects that can be joined and taken apart in your mind. The combinations may be quite unlikely. Start with a simple shape: a cube, a cone, or a sphere. Draw this object. Then draw the same object with another object added to it: a crank handle or another simple shape such as a cylinder. Make a series of drawings in which you add with each new drawing one more shape to the combination you made in the previous drawing; build a fantastic contraption, a Rube Goldbergesque invention, by visualizing and drawing it part by part.

A variant of this exercise it to take the fully realized contraption and disassemble it, pull it apart. Draw the machine, then make separate drawings of the parts as though you were producing an exploded diagram for the owner's manual that accompanies the gadget.

All such drills are beneficial for those working in creative fields. These pushups for the brain exercise and develop the designer's mind to do what it needs to do best: make visual/verbal ideas.

THE ASSOCIATION DRILL

The objective of this drill is to give the brain specific exercise, to ask it to do the kind of work necessary for an on-the-job creative position. This kind of mix-and-match strengthening drill begins with the choice of a general field of interest, a subject with which you are familiar (for two sample drills, I have chosen tennis and children). The next step is to choose one of the major art annual award books—Graphis, The Society of Illustrators Annual, Creativity, and others—and, as you leaf through the book, to let random graphic devices suggest ideas for lead picture-and-word spreads within your chosen subject area.

1. Tennis

In 1981 Bjorn Borg still dominated men's professional tennis; he had won five straight Wimbledon titles and seemed unbeatable. This piece of sculpture, done for the Social Security Administration, suggested a mountain carved with Borg's stern features, all but unclimbable for his challengers.

Jerry Dadds

Rudy Golyn

This poster, which announces a performance of Agatha Christie's play The Mousetrap, *triggered an idea for showing the temptation every nationally ranked young player faces: that of leaving school to get an early start for stardom. The bait is big prize money.*

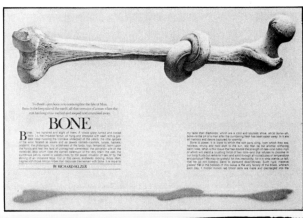

Alan Magee

This fine illustration of a knotted bone, done for a story in Penthouse International, *inspired a lead for a service article that offered advice on how to get unnecessary tension out of your game.*

Howard Sokol

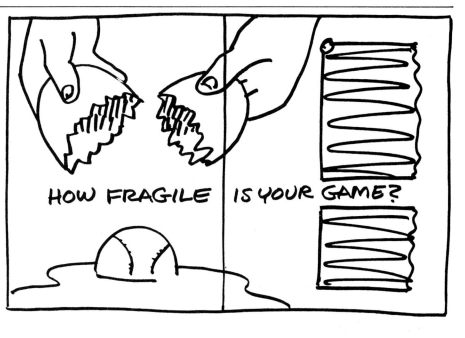

The delicate image of a butterfly being released from an eggshell prompted a lead spread showing a tennis ball yolk falling from a broken ostrich egg, a comment on the ease with which an opponent may upset your predetermined strategy.

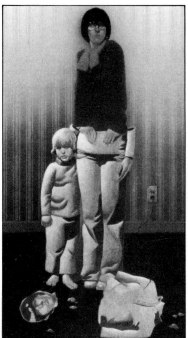

Rick McCollum

Ciba-Geigy ran this illustration of a family in distress on one of its information pamphlets. It triggered the image of a tennis player half-encased in plaster, a symbol of restricted mobility.

The old shell game can be used to talk about disguising your shots. By using the same kind of backswing and shot preparation for all swings you can mislead your opponent. He or she will have minimal time to decide whether you've hit a hard ground stroke, a feathered drop shot near the net, or a passing shot crosscourt.

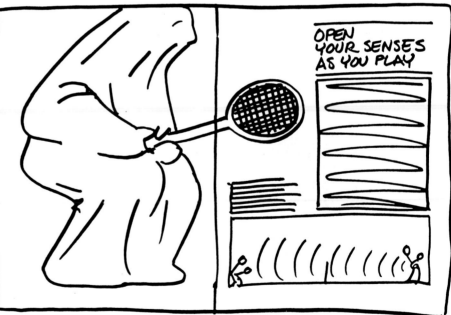

The mysterious shrouded figure in the photograph recalled advice to listen to an opponent's shot as it comes off the racquet. If the sound is loud you can expect a hard, flat shot; if the sound is soft, a short shot is coming and you should probably start immediately for the net.

To date there has been nothing invented—
no vacuum can or repressurizing pump—
to prolong the life of tennis balls. The
image of unbroken eggs in a bottle
suggested a tongue-in-cheek solution to
the problem of preserving tennis balls.

The pen-plumed knight suggested
a player in armor with a crest of
bristling racquets, the person with
a tennis game that's hard to hurt.

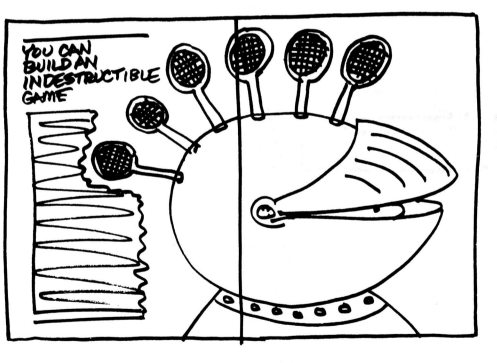

2. Children

This association drill features youth as its subject. Once again, the art annuals are scanned for sources of inspiration to make lead spreads.

Seymour Chwast

Many parents are so anxious for their children to act like adults, to use balanced judgment and to behave responsibly, that the children are denied the benefit of normal childhood fantasies. Pretending games and playacting constitute a healthy part of growing up. The cover showing a burglar taking a painting inspired this illustration of parents actually taking a child's dream away—"for her own good."

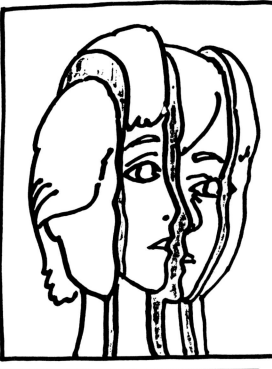

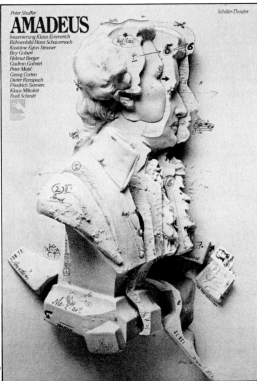

The shattered image of Mozart
on a theater poster prompted an
interpretation of emotional
trauma with this fractured
portrait of a young boy.

The carefree whimsy
of this design studio's
birthday announcement
gave rise to the image
of a boy's dream of
vacation days ahead.

Richard Hess

The suspended bridge in this ad for
Champion Paper products recalled the
uncertainty of teenage years, when it often
seemed that life faded to nothing fore and
aft, that one was stuck indefinitely in the
limbo between childhood and maturity.

Jan Jaromir Aleksiun

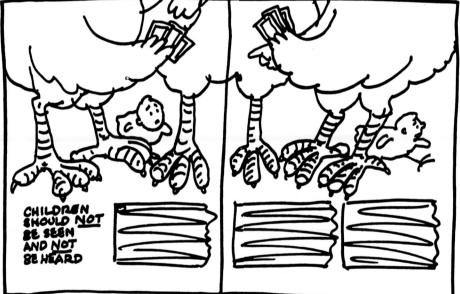

The Anna Livia poster of a woman who is
half chicken prompted the image of a true
hen party. Children are shown as chicks
that are constantly underfoot.

Lauro Giovanetti

Italy's communist party news-paper, l'Unita, commissioned a series of public service posters. This one, a plea for an open and active cultural life in Italy, suggested a story on musical prodigies and their training.

Tomek Kawiak

This poster for a French exhibition of glasswork furnished an image of a broken mirror, the result of uncontrollable anger.

Milton Glaser

Colored light sprays from a glass prism in this poster created for a paint manufacturer. The device became useful to describe the freshness and unorthodoxy with which young children often view life around them.

Jean-Michel Folon

The unraveling head shown on the book jacket suggested a way to peek at the contents of a young boy's mind, to take a look at the teeming inhabitants of an active imagination.

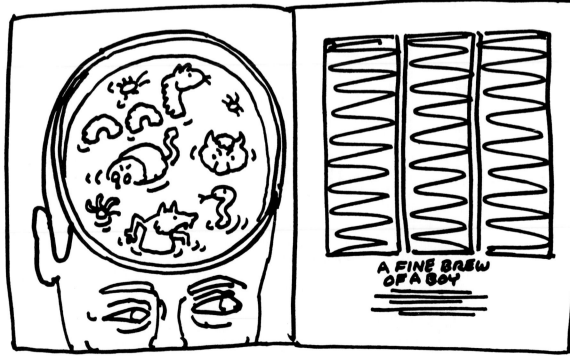

Snips and snails and puppy dog tails — and yoyos and baseball cards and pet frogs — these are what little boys are made of. The trigger image is a poster for a Columbia University crafts festival.

A line of lightbulb-headed chorus girls yielded a story on the teaching of gifted children who sit in classrooms with other children of average ability.

ADDING AND SUBTRACTING IN THE MIND

This two-part exercise developed by Bruce Bacon has value for all creative professionals. The first part consists of adding elements in a series of drawings so that a simple shape grows to ever greater complexity as other parts are fit onto it. The other half of the drill starts with a fantastic construction that is then disassembled and sorted into separate component parts.

This spread shows one example of each kind of exercise Bacon recommends to his design students. At right is an additive series of drawings. Bacon starts with a cube and installs a random assortment of seemingly unrelated objects. The final machine gains its own kind of integrity because the parts appear to fit. On the opposite page a fantastic construction is taken apart visually.

Bruce Bacon

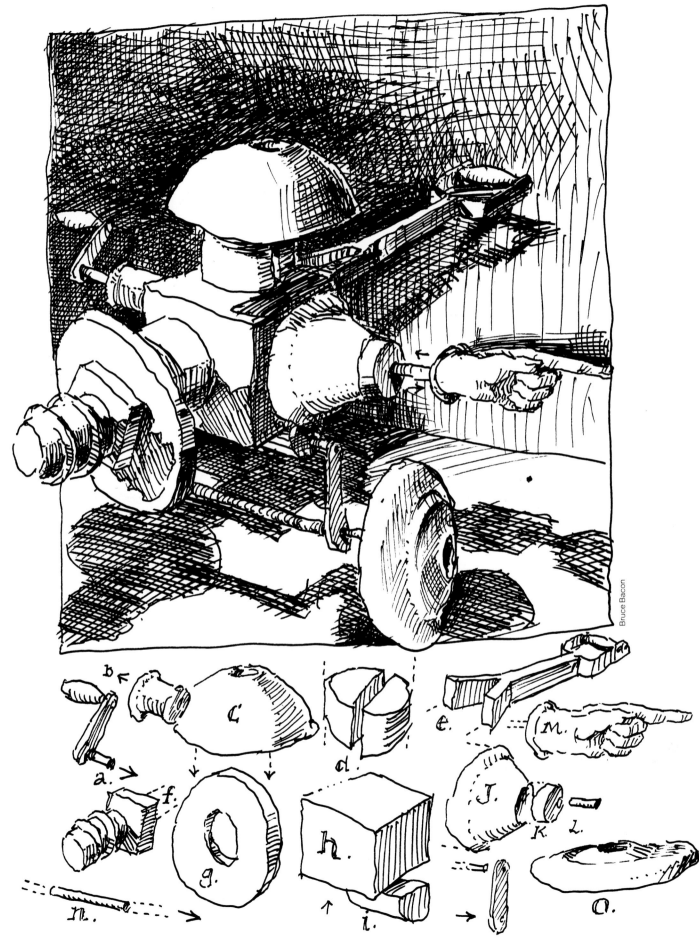

Bruce Bacon

This page shows another example of the additive drill. The same kind of object-in-transition treatment is used in a more purposeful and disciplined way by Elwood Smith. His cover, on the opposite page, follows the mental rise and fall of a hero, a meek little soul who grows to meet an imagined challenge and then sinks back to, well, nothing.

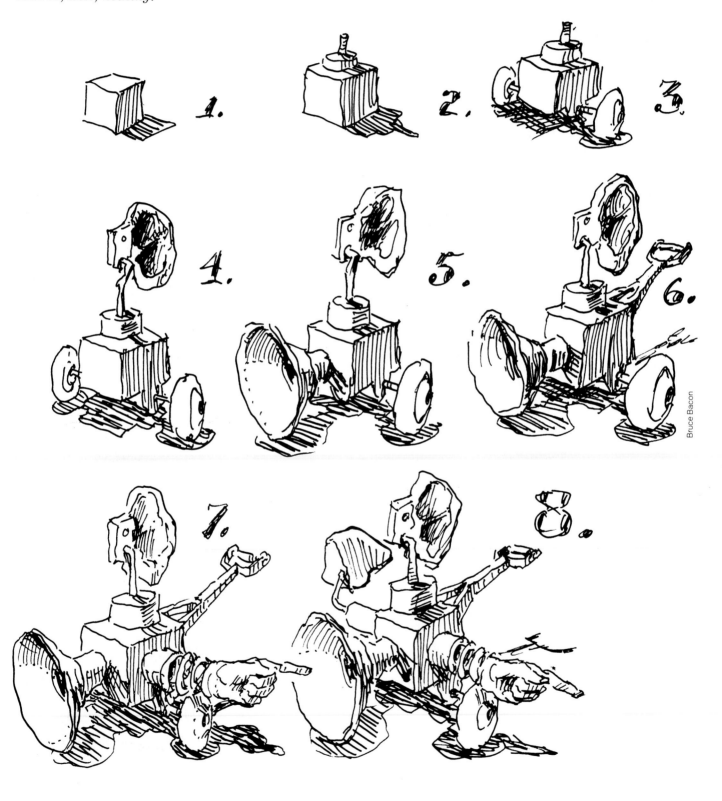

Bruce Bacon

Elwood Smith

MLO

MEDICAL LABORATORY OBSERVER • MARCH 1983

The impact of nonrevenue-producing activities

A part-time MT degree program

How to lead a problem-solving meeting

Caution on the slippery road to competitive bidding

COST CONTAINMENT
THE SQUEEZE GETS TIGHTER
TACTICS FOR THE 80s

Walter Wick (art by Judith Jampel)

LOW BUDGETS: CHAMPAGNE DESIGN WITH BEER MONEY

The dollar sign is repeatedly pulled into symbolic service. What can you do with a dollar? The following pages will show a few solutions.

You can squeeze it . . .

The subject of this cover is managing operating expenses in a medical laboratory. The symbol, made of rubber foam and covered with a green polyester fabric, is constricted by a bright blue vinyl belt.

Competition at different career levels of professional art has different requirements. In some ways competing is harder in the early years—but this is not entirely for reasons of inexperience. A special set of difficulties revolves around the art budget, the amount of money allocated by the publisher to buy illustration and photography.

Many of you who read this will be working for publications with little or no money available for dramatic illustration or photography. You may be muttering, "Sure, the big graphics and special effects you show in this book are wonderful, but these things have little to do with me. I don't have the luxury of a huge art budget to do my covers and pages."

First, many of the examples of eye-catching design and "impossible" photographs were not as expensive to do as you may think. There are ways to reduce cost and produce surprisingly impressive pieces of graphic display by sheer hustle—by using beginning artists and photographers, by buying pieces of photography directly out of a photographer's portfolio when you see an upcoming article they might adorn, by buying second reproduction rights to art that runs in national publications or in regional magazines from other parts of the country, by tapping design and illustration classes in universities, by making collages with reproducible art that is free of copyright restriction, or by using display typography. (If there is very little illustration and few or no ads in your publication, display typography—headlines, quotes, and initials—can be an effective art form in itself.)

You can cut the cost of photographs considerably if you spend the time and energy to scrounge your own props; you may even find the means to set up your own simple pocket photo studio. You can train yourself to take very adequate photographs; a designer's eye can be quite useful when it looks through a camera viewfinder. (Many top photographers who have been art directors would agree, people such as Carl Fischer, Art Kane, Bill Cadge, and Henry Wolf.) Learning the mechanics of a good medium-priced 35mm camera is not difficult, either. One of the masters of low-cost editorial design is Greg Paul, former art director of the *Cleveland Plain Dealer* Sunday magazine. He has produced good graphics on the proverbial shoestring and has gladly accepted the challenge of having to do so month after month.

What should be emphasized, however, is that developing the skill to create lively ideas can put sparkle in your publication by itself. If you can produce a good concept for your magazine cover, you can find a less expensive way to get the idea executed and still make a modest graphic splash; you can still get your book noticed. Good ideas realized in a mediocre way are still far better than mundane ideas executed with technical skill and polish. The quality of thinking evidenced by such work may be the bright young designer's passport to a better job on a publication with more budgetary freedom. Low budgets are explicable to a potential employer; numb or careless communication is not.

Jack Lefkowitz, art director of *Industrial Launderer*, puts it this way: "Make sure you have ideas that carry themselves no matter how simple the technique is. If you need technique to make an illustration interesting, then you're missing something in the idea. If your idea is strong enough, especially if the concept of the illustration is worked out in tandem with the editorial material, it becomes unimportant whether you've done your drawing in ink or pencil or paint or whatever. The concept is very, very important. It's more important than style."*

The purpose of this book is to help the young artist find the first solid rungs of the ladder he or she must

Folio, April 1984: 24.

... freeze it ...

A cover image for a magazine aimed at pharmacists alluded to third-party (delayed) insurance payments for drugs. The article talks about ways to unfreeze reimbursement funds.

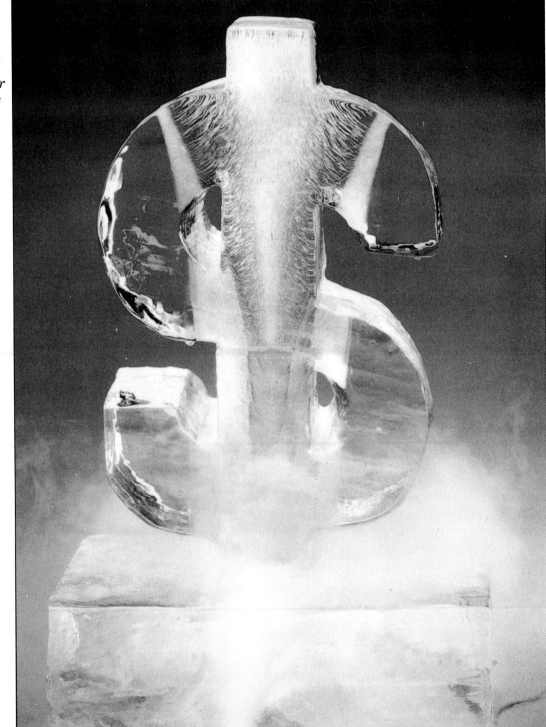

Walter Wick

climb, to furnish objectives and pretested ways to reach them. None of the methods described requires brilliance—only resolution and effort. But the suggestions these pages offer can be of real help in what has been a landscape devoid of signposts.

This book will not make you a Milton Glaser or a George Lois, but it will point the way to substantial improvement of your skills. Professional communication is complex, demanding work; perhaps the surest way to reach the top levels is to become an apprentice or associate of one of the best in the business.

But there is much you can do to improve your lie (as a golfer might say), to equip yourself better to attain the position that will allow you to do what you really want to do. The Bite System is a crude tool with which to start a supremely delicate task, but it is a tool. It can be used by anyone, and it does get results. It will probably be improved or supplanted by more efficient systems as theory grows in this neglected area of technique, but, for the time being, it is the only tool extant to help get a graphic design job done.

As you use the Bite System, you will get so familiar with its parts that you won't have to jot down the variations and can list the choices in your mind: you will

... burn it ...
The article described a pension fraud. A group of doctors allowed a fast-talking salesman to utilize their collective savings for a get-rich-quick deal. Poof, no money.

Stephen E. Munz (art by David Cayce)

turn over the options in your head and go directly to doodled visuals. And, in those unusual cases where a problem stubbornly defies solution, it is simple to go back to making lists to expose new ideas.

The conditioning exercises are important, too. There is no line of gainful employment that does not require hard work and dedication if the necessary skills and knowledge are to be gotten. If you wish to get your share of work, you must compete successfully with others who are equally determined and have strong abilities of their own. The overall goal of this book is to help you compete.

The hope is that you will return to your work with a new sense of excitement, a new awareness of potential. This profession offers satisfactions that are all too rare in this increasingly mechanized world. Graphic design is an intensely human kind of work, perhaps one of the last shelters for the Renaissance Man or Woman. It offers samplings of the whole range of human industry and, by its nature, forbids routine; variety and change are its coin. A philosopher once said, "For most men life is one unrelenting search for the right envelopes in which to file themselves." Creative people are not of this breed—they are very much alive.

John Trotta

. . . age it . . .
Investment advice covering portfolio action to be spread throughout the calendar year sported this image of a fund for all seasons.

. . . surround it . . .
This cover article offered a detailed checklist for monitoring and controlling small routine costs in the laboratory.

Walter Wick

Stanislaw Zagorski

. . . or idealize it.
*A financial service
commissioned this vivid
Magritte-like dollar
in the sky illustration.*

PICTURE CREDITS

Cover: poster for Imhara and Keep Advertising, Inc. of Santa Clara, California. Agency: Philip Bauer, Inc. Photographer: Carol Sollecito.

PREFACE

vi: self-promotion photo by Carl Flatow.
viii: poster for the Edison Electric Institute. Agency: Nerco, Inc.
ix: self-promotion photo by Parish Kohanim.

INTRODUCTION

x: cover for house organ, *Batesworld*. Agency: Ted Bates Worldwide. Photographer: Bjorn Winsnes.
x: poster for sales staff of The Manville Corporation. Designer: Bill Bechtel.
xi: self-promotion photo by Bruno R. Shreck.

WHY IS THIS BOOK NEEDED?

xii: cover for *Pushpin Graphic*, 1980. Illustrator: Elwood Smith.
4: advertisement for Adidas running shoes. Agency: Laurence, Charles & Free, Inc. Art director: Anne Occi. Illustrator: Wayne McLoughlin.
5: advertisement for Adidas tennis shoes. Agency: Laurence, Charles & Free, Inc. Art director: Brett Shevack. Illustrator: Willardson and White Studios.
6: cover for *Texas Monthly* magazine. Art director: Fred Woodward. Photographer: Robert Latorre. Artist: Joseph Melancon.
7: article on burnout for *Medical Economics* magazine. Art director: Barbara Groenteman. Photographer: Stephen E. Munz.
7: advertisements for Baccarat glassware. Agency: Young & Rubicam, Inc. Creative director: John Geoghegan. Art director: John Martinez. Photographer: Andrew Unangst.
8: album covers designed by Paula Scher for CBS Records. "Multiplication" illustration by David Wilcox, "Ear Meal" illustration by Richard Hess and "Bad Luck" illustration by David Wilcox.
9: advertisement for Pounce insecticide. Agency: Marsteller, Inc. Art director: Harry Kerker. Illustrator: Bob Krogel.
9: article about jet lag from *Psychology Today* magazine. Art director: Carveth Hilton Kramer. Illustrator: Robert Grossman.
10: cover about diet colas for *New York* magazine. Art director: Walter Bernard. Photographer: Ben Somoroff.
10: article about jealous love from *Psychology Today* magazine. Art director: Carveth Hilton Kramer. Illustrator: Marvin Mattelson.
11: cover about dropped-out executives for *Forbes* magazine. Art director: Everett Halverson. Illustrator: Richard Mantel.
11: cover about trimming costs for *Medical Economics* magazine. Art director: Barbara Groenteman. Photographer: Stephen E. Munz.
11: advertisement for Kitty Litter brand box fill. Agency: J. Walter Thompson. Creative director: David DeVary. Art director: Andy Anema. Photographer: Dennis Manarchy.
12: advertisement for Congoleum floor covering. Photographer: Mark Kozlowski.
13: cover about car manufacturing for *Business Week* magazine (McGraw-Hill, Inc.). Art director: Malcolm Frouman. Illustrator: Braldt Bralds.
14: cover about Japanese business for *Forbes* magazine. Art director: Everett Halverson. Illustrator: Kinuko Craft.
15: article about accountants for *Medical Economics* magazine. Art director: Barbara Groenteman. Photographer: Walter Wick.
15: article about managers and inventors for *Innovation* magazine. Art director: Eric Gluckman. Illustrator: John Newcomb.
16: covers for *Medical Laboratory Observer* magazine. Art director: Tom Darnsteadt. Photographer: Stephen E. Munz. Artist for "burnout" cover: Janis Conklin.
16: article about saving money for retirement for *Medical Economics* magazine. Art director: Barbara Groenteman. Illustrator: Lonni Sue Johnson.
17: advertisements for Timken Steel products. Agency: Batten, Barton, Durstine and Osborn. Creative director: Terry Scullin. Art director: Ruthann Richert.
18–19: posters for the Milwaukee Museum of Art. Agency: Frankenberry, Laughlin and Constable. Creative director: Dennis Frankenberry. Art director: Rachel Stephens.

Photographers: Peter Amft and Don Wilson.
19: brochure cover for The Communicating Arts Group of Arizona. Agency: Phillips and Ramsey. Art director: Dennis Merritt. Photographer: Rick Gayle.

THE BITE SYSTEM: A RELIABLE STARTING POINT

21,22: article about investment for *Medical Economics* magazine. Art director: Barbara Groenteman. Photographer: Walter Wick.
25–45: all demonstration sketches by John Newcomb.
45: article about tennis cheaters for *Tennis* magazine (The New York Times Co.). Art director: Michael Brent. Photographer: Jeffrey Fox.
46–86: all sketches by John Newcomb.
47: cover about inflation for *Medical Economics* magazine. Art director: Barbara Groenteman. Photographer: Walter Wick.
49: cover about hospital chains for *Medical Economics* magazine. Art director: Barbara Groenteman. Photographer: Walter Wick.
51: cover about picking investment winners in an election year for *Medical Economics* magazine. Art director: Barbara Groenteman. Photographer: Walter Wick. Artist: Kathy Jeffers.
53: cover about financial strategies for *Medical Economics* magazine. Art director: Barbara Groenteman. Photographer: Stephen E. Munz. Artist: Al Pisano.
55: cover about the resale value of cars for *Medical Economics* magazine. Art director: Barbara Groenteman. Photographer: Stephen E. Munz. Artist: Kathy Jeffers.
57: cover about doctor policing for *Medical Economics* magazine. Art director: Barbara Groenteman. Photographer: Walter Wick.
59: cover about a malpractice suit for *Medical Economics* magazine. Art director: Barbara Groenteman. Photographer: Stephen E. Munz.
61: cover about wasting tax dollars for *Medical Economics* magazine. Art director: Barbara Groenteman. Photographer: Walter Wick. Artist: The Manhattan Model Works.
63: cover about tax audits for *Medical Economics* magazine.

Art director: Barbara Groenteman. Photographer: Stephen E. Munz. Artist: Asdur Takakjian.
65: cover about 24-hour clinics for *Medical Economics* magazine. Art director: Barbara Groenteman. Photographer: Stephen E. Munz. Artist: Kathy Jeffers.
67: cover about young doctors for *Medical Economics* magazine. Art director: Barbara Groenteman. Photographer: Walter Wick.
69: cover about PPO's for *Medical Economics* magazine. Art director: Barbara Groenteman. Photographer: Walter Wick.
71: cover about long-term investment winners and losers for *Medical Economics* magazine. Art director: Barbara Groenteman. Photographer: Walter Wick.
73: cover about same-day surgery for *Medical Economics for Surgeons*. Art director: William Kuhn. Photographer: Stephen E. Munz.
75: cover about competition for *Medical Economics for Surgeons*. Art director: William Kuhn. Photographer: Stephen E. Munz.
77: cover about medicine in 1990 for *Medical Economics* magazine. Art director: Barbara Groenteman. Illustrator: Lonni Sue Johnson.
79: cover about peer review for *Medical Economics* magazine. Art director: Barbara Groenteman. Photographer: Stephen E. Munz. Artist: Margaret Garrison.
81: cover about cocaine addiction for *Medical Economics* magazine. Art director: Barbara Groenteman. Photographer: Stephen E. Munz.
83: cover about government regulation for *Medical Economics* magazine. Art director: Barbara Groenteman. Photographer: Walter Wick.
85: cover on earnings for *Medical Economics for Surgeons*. Art director: William Kuhn. Photographer: Stephen E. Munz.
87: cover about rising malpractice rates for *Medical Economics for Surgeons*. Art director: William Kuhn. Photographer: Walter Wick.

SHAPING YOUR CONVERSATION WITH THE READER

88: article on partnerships for *Medical Economics* magazine.

Art director: Barbara Groenteman. Photographer: Stephen E. Munz. Artist: The Manhattan Model Works.
91: cover on the Annual Bum Steer Awards for *Texas Monthly* magazine. Art director: Fred Woodward. Illustrator: Tom Curry.
91: cover about a card game puzzle for *Games* magazine (Playboy Enterprises, Inc.). Art director: Don Wright. Photographer: Walter Wick. Artist: Sandra Forrest.
91: cover about magicians for *Geo* magazine (Knapp Communications Corp.). Art director: Mary Kay Baumann. Photographer: Chris Callis.
93: article about medical consent for *Medical Economics for Surgeons*. Art director: William Kuhn. Illustrator: Andrea Baruffi.
94: article about Brazilian natives for *Geo* magazine (Knapp Communications Corp.). Art director: Mary Kay Baumann. Photographers: Maureen Bisilliat and Jean-Pierre Dutilleux.
94: article about video game designers for *Geo* magazine (Knapp Communications Corp.). Art director: Mary Kay Baumann. Photographer: Mark Hanauer.
95: article about estimating distance for *Golf Digest* magazine (The New York Times Co.). Art director: John Newcomb. Illustrator: Elmer Wexler.
96: article about a fictional tryst for *Audience* magazine. Art director: Seymour Chwast. Illustrator: Seymour Chwast.
97: article about a California legal battle for *Medical Economics* magazine. Art director: Barbara Groenteman. Photographer: Stephen E. Munz. Artist: Kathy Jeffers.
97: article about the popularity of tennis for *Tennis* magazine (The New York Times Co.). Art director: Michael Brent. Photographer: Mark Kozlowski.
98: article about overtraining for *Medical Economics* magazine. Art director: Barbara Groenteman. Illustrator: Daniel Maffia.
99: article about castles in Spain for *Geo* magazine (Knapp Communications Corp.). Art director: Mary Kay Baumann. Photographer: Reinhart Wolf.
99: article about Firestone Country Club in Akron, Ohio for *Golf Digest* magazine (The New York Times Co.). Art director:

John Newcomb. Photographer: John Newcomb.
100: from a portfolio in *American Photographer* magazine (CBS Publications), two photos by Galen Rowell. Art director: Will Hopkins.
101: title page from *A Bridge Too Far*, a special publication of the New York Times. Art director: John Newcomb. Photographer: Robert Penn.
101: article about personal finance for *Medical Economics* magazine. Art director: Barbara Groenteman. Photographer: Winfred Meyer.
102: masthead spread from *A Bridge Too Far*, a publication of the New York Times. Art director: John Newcomb. Photographer: Robert Penn.
104: self-promotion piece photographed by Dennis Chalkin. The acrobatic model is Michael Heintz.
105: news coverage of a summer festival in the South. Photographer: Chip Henderson.
106: article about the Rolls Royce for *Medical Economics* magazine. Art director: Barbara Groenteman. Photographer: Anthony Vaccaro.
106: article about a champion doubles team for *Tennis* magazine (The New York Times Co.). Art director: John Newcomb. Photographer: John Newcomb.
107: article about salesmen for *Medical Economics* magazine. Art director: Barbara Groenteman. Photographer: Stephen E. Munz. Artist: Asdur Takakjian.
107: brochure for an ammunition company, The Patton and Morgan Corporation, Inc. Art director: John Newcomb. Photographer: Stephen Szurlej.
108: article about John Belushi's movie *1941* for *Rolling Stone* magazine (Straight Arrow Publications, Inc.). Art director: Mary Shanahan. Photographer: Bonnie Schiffman.
110: article about animals of the Serengeti for *Geo* magazine (Knapp Communications Corp.). Art director: Mary Kay Baumann. Photographer: Reinhard Kuenkel.
110: article about Jack Nicklaus for *Golf Digest* magazine (The New York Times Co.). Art director: John Newcomb. Photographer: John Newcomb.
111: a series of articles about giving physical examinations for *RN* magazine. Art director: Hector Marrero. Photographer: Stephen E. Munz.

112: article about Nazi fugitives for *Life* magazine (Time, Inc.). Art director: Bob Ciano. Photographers: (news picture) The Yad Vashem Museum, (witness) David Rubinger and (the closeup of Frank Walus) Enrico Ferorelli.
114: opening page for a big-prize-money section in *Golf Digest* magazine (The New York Times Co.).
114: article about a retired golfer for *Golf Digest* magazine (The New York Times Co.). Art director: John Newcomb. Photographer: Leonard Kamsler.
115: section opener for a series of articles about Sam Snead in *Golf Digest* magazine (The New York Times Co.). Art director: John Newcomb. Illustrator: Stan Drake. Photographer: Leonard Kamsler.
116: article about camping gear for *Atlanta Weekly* magazine *(The Atlanta Journal and Constitution)*. Art director: Ike Hussey. Photographer: Ed C. Thompson.
116: article about financial disaster for *Medical Economics* magazine. Art director: Barbara Groenteman. Illustrator: Lowell Hess.
117: article about space technology and business for *Geo* magazine (Knapp Communications Corp.). Art director: Mary Kay Baumann. Photographer: Norman Seeff.
118: "Bum Steer Award" cover of *Texas Monthly* magazine. Art director: Fred Woodward. Illustrator: Tom Curry.
119: cover about WW II airplanes for *True* magazine (Fawcett Publications, Inc.). Art director: John Newcomb. Photographer: Ray Woolfe.
119: article about goalie conditioning for *Hockey* magazine (The New York Times Co.). Art director: Laura Duggan. Photographer: John Newcomb.
120: article about the Washington National Zoo for *Attenzione* magazine. Art director: Patricia Nordin. Photographer: Steve Brown.
121: article about Ion Tiriac for *Tennis* magazine (The New York Times Co.). Art director: Stan Braverman. Photographer: John Newcomb.
121: article about OPEC oil for *Medical Economics* magazine. Art director: Barbara Groenteman. Photographer: Stephen E. Munz.
122: cover about Vic Braden for *Tennis* magazine (The New

York Times Co.). Art director: Michael Brent. Photographer: Mark Kozlowski.
123: cover about watching routine costs for *Medical Laboratory Observer* magazine. Art director: Tom Darnsteadt. Illustrator: Barnett Plotkin.
124: opening page for a section on putting for *Golf Digest* magazine (The New York Times Co.). Art director: John Newcomb. Photographer: Ralph Breswitz.
124: article about legal risk for *Medical Economics* magazine. Art director: Barbara Groenteman. Photographer: Stephen E. Munz.
125: cover about balance for *Golf Digest* magazine (The New York Times Co.). Art director: Pete Libby. Photographer: John Newcomb.
126: article about near-drowning for *RN* magazine. Art director: Hector Marrero. Photographer: Shig Ikeda.
127: cover about retirement notice for *Medical Economics for Surgeons*. Art director: William Kuhn. Photographer: Walter Wick.
128: article about famine in Uganda for *Life* magazine (Time, Inc.). (hands) Mike Wells/Aspect Picture Library, (bed) Michel Folco/Gamma-Liaison Agency.
130: article about Juan Rodriguez for *Golf Digest* magazine (The New York Times Co.). Art director: John Newcomb. Photographer: John Newcomb.
132: article about ambitious young golfers for *Golf Digest* magazine. (The New York Times Co.). Art director: John Newcomb. Photographer: stock.
132: article about four seasons of golf for *Golf Digest* magazine. (The New York Times Co.). Art director: Pete Libby. Photographer: Richard Beattie.
134: cover about screening procedures for patients for *Diagnosis* magazine. Art director: Tim McKeen. Photographer: Bill Longcore.
135: cover about the causes of fainting for *Diagnosis* magazine. Art director: Tim McKeen. Photographer: Walter Wick.
136: article about ambitious young executives for *Innovation* magazine. Art director: Eric Gluckman. Illustrator: John Newcomb.
137: illustration for a poem, "Jimmy's got a Goil," in *Pushpin Graphic* magazine. Art

director: Seymour Chwast. Illustrator: Elwood Smith.
138: article about finding models, "The Playmate Process," in *Audience* magazine. Art director: Seymour Chwast. Illustrator: Robert Grossman.
140: article on professional humor, "The Gag Writers," in *Audience* magazine. Art director: Seymour Chwast. Illustrators: (from left): Daniel Maffia, Philip Hays, Gilbert Stone, and Seymour Chwast.
142: article on sand play in *Golf Digest* magazine (The New York Times Co.). Art director: John Newcomb. Illustrator: James McQueen.
143: article about locker room atmosphere at golf tournaments in *Golf Digest* magazine (The New York Times Co.). Art director: John Newcomb. Artist: Robert Heindel.
144: special report, "Living Single, Sleeping Double," in *Rolling Stone* magazine (Straight Arrow Publications, Inc.). Art director: Mary Shanahan. Illustrator: Lib Cummings.
145: profile of three taxidermist brothers for *True* magazine (Fawcett Publications, Inc.). Art director: John Newcomb. Photographer: Daniel Kramer.
146: article on therapeutic drug monitoring in the laboratory for *Medical Laboratory Observer* magazine. Art director: Tom Darnsteadt. Photographer: Walter Wick.
147: cover about choosing the best-qualified member of your staff for promotion for *Medical Laboratory Observer* magazine. Art director: Roger Dowd. Photographer: Stephen E. Munz. Artist: Barbara Slocum.
148: article about careless record-keeping for *Medical Economics* magazine. Art director: Barbara Groenteman. Photographer: Stephen E. Munz.
149: article about a psychotic patient for *Medical Economics* magazine. Art director: Barbara Groenteman. Photographer: Stephen E. Munz. Artists: Kathy Jeffers (head) and Margaret Cusack (body).
150: article about military waste in spending for *Medical Economics* magazine. Art director: Barbara Groenteman. Photographer: Stephen E. Munz. Artist: Janis Conklin.
151: article about insensitive doctors for *Medical Economics* magazine. Art director:

Barbara Groenteman. Photographer: Stephen E. Munz.
152: article on the medicinal uses of marijuana for *Psychology Today* magazine. Art director: Carveth Hilton Kramer. Photographer: Jim Houghton.
153: article about interviewing guidelines for *Medical Laboratory Observer* magazine. Art director: Tom Darnsteadt. Photographer: Walter Wick. Artist: The Manhattan Model Works.
154: cover about using light-weight clubs for *Golf Digest* magazine (The New York Times Co.). Art director: John Newcomb. Photographer: Bill Strode. Artist: Bruce Roggieri.
154: article about new IRS tax rules for *Medical Economics* magazine. Art director: Barbara Groenteman. Photographer: Stephen E. Munz.
155: article about attracting patients for *Medical Economics* magazine. Art director: Barbara Groenteman. Photographer: Ken Schroers. Artist: Asdur Takakjian.
156: article about the struggle between fans of the rubber-covered golf ball and the devotees of plastic ball covers for *Golf Digest* magazine (The New York Times Co.). Art director: John Newcomb. Photographer and artist: Jerry Cosgrove.
156: article about the alligator gar for *True* magazine (Fawcett Publications, Inc.). Art director: John Newcomb. Illustrator: Jerry Cosgrove.
158: article about investing in farm land in *Medical Economics* magazine. Art director: Barbara Groenteman. Photographer: Stephen E. Munz.
159: article about a doctor who tried to oversee the construction of his house in *Medical Economics* magazine. Art director: Barbara Groenteman. Photographer: Stephen E. Munz. Artist: Janis Conklin.
160: self-promotion photograph of backfiring pistol by Angus Forbes.
161: article about baseball enthusiasm for *Playboy* magazine. Art director: Tom Staebler. Photographer: Richard Izui. Artist: Parviz Sadighian.
164: cartoon strip from *Adweek* magazine by Stan Mack. Art director: Walter Bernard.

TRAIN YOURSELF TO BE AN EFFECTIVE DESIGN THINKER

166: self-promotion drawing of a designer switching heads by Stan Mack.

174: photo of George Lois by John Newcomb.

175: all covers of *Esquire* magazine photographed by Carl Fischer. Art director: George Lois.

176: cover of *Ads* magazine photographed by Carl Fischer. Art director: George Lois.

177: covers of *Cuisine* magazine (CBS Publications, Inc.): James Beard photographed by Carl Fischer, Dagwood courtesy of King Features Syndicate, Inc. Art director: George Lois.

177: photo of George Lois by John Newcomb.

178: photo of Stan Mack by John Newcomb.

178: illustration of little man in knight suit by Stan Mack.

179: cartoon strips from *Adweek* magazine by Stan Mack. Art director: Walter Bernard.

180: photo of Guy Billout by John Newcomb.

180: self-promotion card by Guy Billout.

181: cover about capital punishment for *The New Republic* magazine. Illustrator: Guy Billout.

181: advertisement for Bruno Dessange, a hair styling salon opening in New York. Art director: Craig Braun. Illustrator: Guy Billout.

182: illustration of Noah's liner for *Squid and Spider, A Look at the Animal Kingdom* (Prentice-Hall, Inc.). Art director: Barbara Francis. Illustrator: Guy Billout.

183: illustration of mountain climber and spider for *Squid and Spider, A Look at the Animal Kingdom* (Prentice-Hall, Inc.). Art director: Barbara Francis. Illustrator: Guy Billout.

183: photo of Guy Billout by John Newcomb.

184: drawings for *Avenue* magazine by Guy Billout. Art director: Ray Hooper.

184: illustration of Loch Ness monster and lighthouse for *Stone and Steel: A Look at Engineering* (Prentice-Hall, Inc.). Illustrator: Guy Billout.

185: illustration of lion and zebra on a suburban highway for *Atlantic Monthly* magazine.

Art director: Judy Garlan. Illustrator: Guy Billout.

186: photo of Milton Glaser by John Newcomb.

186: poster for the American Film Institute on the National Video Festival. Illustrator: Milton Glaser.

187: poster for The New York Zoological Society for a special film event, "The Night of the Snow Leopard." Illustrator: Milton Glaser.

188: planning sketches and finished poster for the American Institute of Graphic Arts. Illustrator: Milton Glaser.

189: photo of Milton Glaser by John Newcomb.

190: details from a calendar for Zanders feinpapiere A.G., a European printer. Illustrator: Milton Glaser.

191: photo of Milton Glaser by John Newcomb.

192: photo of Seymour Chwast by John Newcomb.

192: poster for the Art Institute of Fort Lauderdale. Art director: Edward A. Hamilton. Illustrator: Seymour Chwast.

193: cover for a special section of *Adweek* magazine on creativity. Art director: Walter Bernard. Illustrator: Seymour Chwast.

193: poster for a 1982 peace march in New York. Executed for the June 12 Rally Committee of New York. Illustrator: Seymour Chwast.

194: poster for the British Broadcasting Company's presentation of "I, Claudius." Illustrator: Seymour Chwast.

194: cover for a writers' guidebook to New York markets. Produced for the New York City Department of Cultural Affairs. Art director: Toshiake Ide. Illustrator: Seymour Chwast.

194: photo of Seymour Chwast by John Newcomb.

195: photo of Seymour Chwast by John Newcomb.

195: magazine advertisements for *Forbes* magazine. Art director: Warren Westwood. Illustrator: Seymour Chwast.

196: photo of Paula Scher by John Newcomb.

196: album cover, "Nice to Have Met You," for CBS Records, Inc. Art director: Paula Scher. Illustrator: Richard Hess.

197: album cover, "Rush Hour," for CBS Records, Inc. Illustrator: Robert Giusti.

198: album cover, "Heart of the City," for Atlantic Records, Inc. Photographer: Arnold

Rosenberg.

198: album cover, "Yardbirds Favorites," for Epic Records, Inc. Illustrator: David Wilcox.

199: photo of Paula Scher by John Newcomb.

199: three album covers produced for Tappan Zee Records, Inc. Photographer: John Paul Endress.

200: photo of Elwood Smith and Eleanor by John Newcomb.

200: cover, "Man's Best Friend," for the *Chicago Tribune Magazine*. Art director: Dan Jursa. Illustrator: Elwood Smith.

201: detail from a wall bulletin, *Pet Care Report*, published by the 13-30 Corporation, Inc. Art director: Mike Marcum. Illustrator: Elwood Smith.

202: cover for a special section about new software companies for *Datamation* magazine. Art director: Ken Surabian. Illustrator: Elwood Smith.

202: study sketch of artist on luge by Elwood Smith.

203: cover on Christmas theme for the *Chicago Tribune Magazine*. Art director: Dan Jursa. Illustrator: Elwood Smith.

203: cover, "The Spirit of Christmas Past," for *Westward* magazine (The Dallas Times Herald Corporation). Art director: James Smith. Illustrator: Elwood Smith.

204: spread on New York City in the next interglacial period for *Vanity Fair* magazine (Condé Nast Publications, Inc.) Art director: Lloyd Ziff. Illustrator: Elwood Smith.

204: cover "Hot Diggety Dogs!" for the *Chicago Tribune Magazine*. Art director: Dan Jursa. Illustrator: Elwood Smith.

205: article about adhesives for *Technology Today* magazine. Art director: Noel Werritt. Illustrator: Elwood Smith.

205: photo of Elwood Smith by John Newcomb.

206: article about the development of advertising slogans for *Adweek* magazine. Art director: Walter Bernard. Illustrator: Elwood Smith.

206: notice of relocation by Elwood Smith.

208: photos of Anthony Angotti by John Newcomb.

208: advertisements for BMW automobiles. Agency: Ammirati & Puris, Inc. Art director: Anthony Angotti. Photograph of automobile by Elliot Resnick.

209: television commercial for Club Med Resorts, Inc. Agency:

Ammirati & Puris, Inc. Art director: Anthony Angotti. Photographer: Michael Sarasin.

209: photo of Anthony Angotti by John Newcomb.

210: photos of Nancy Rice and Tom McElligott by Judy Carter.

210: advertisement for WTCN–TV, "Use drugs..." Agency: Fallon McElligott Rice, Inc. Art director: Pat Burnham. Photographer: Craig Perman.

211: advertisement for WTCN–TV, "Their love is undying..." Agency: Fallon McElligott Rice, Inc. Art director: Pat Burnham. Photographer: Craig Perman.

211: photo of Tom McElligott by Judy Carter.

211: two posters for the Minnesota Nuclear Freeze Campaign. Agency: Fallon McElligott Rice, Inc. "If this is what happens to a doll...": art director, Nancy Rice; photographer, Kerry Petersen/Marvy. "Sure the world can survive...": art director, Nancy Rice; photograph, The Bettman Archive.

212–214: advertisements and posters for the Episcopal Church. Agency: Fallon McElligot Rice, Inc. "...His only begotten Son...": art director, Nancy Rice; photographer, Tom Bach/Marvy. "If Jesus fed the multitudes...": art director, Nancy Rice; photographer, Tom Bach/Marvy. "...regardless of...the number of times you've been born.": art director, Nancy Rice. "...a religious experience without hallucinations...": art director, Nancy Rice; photographer, Jim Arndt/Berthiaume. "If you think it's inconvenient being a Christian...": art director, Nancy Rice. "Whose birthday is it?": art director, Nancy Rice. "...Jesus had his doubts...": art director, Nancy Rice.

215: poster for Knox hardware stores, "Put your husband on a pedestal." Agency: Fallon McElligott Rice, Inc. Art director: Pat Burnham. Photographer: Frank Miller.

215: advertisements for ITT Life Insurance Corporation. Agency: Fallon McElligott Rice, Inc. "Your whole life is a mistake": art director, Dean Hanson; illustrator, Scott Baker. "...a beautiful body should be worth more...": art director, Nancy Rice; photographer, Dick Jones.

216: photo of Bruce Bacon by John Newcomb.

216–219: all paintings, still life photographs and sculpture by Bruce Bacon.

219: photo of Bruce Bacon by John Newcomb.

220–223: all photos of Dik Browne and his family and all cartoon strips by courtesy of King Features Syndicate, Inc.

224: photo of Carveth Hilton Kramer by John Newcomb.

224: sketch of dentist and electric chair by Carveth Hilton Kramer.

225: cover about inadvertent mistakes for *Psychology Today* magazine. Art director: Carveth Hilton Kramer. Illustrator: Robert Grossman.

225: sketch by Carveth Hilton Kramer and illustration for story, "Fear of Skiing," for *Psychology Today* magazine. Art director: Carveth Hilton Kramer. Illustrator: Seymour Chwast.

226: sketch by Carveth Hilton Kramer and cover about dream research for *Psychology Today* magazine. Art director: Carveth Hilton Kramer. Illustrator: Marvin Mattelson.

227: article about keeping secrets for *Psychology Today* magazine. Art director: Carveth Hilton Kramer. Photographer: Mark Kozlowski.

233–243: all sketches by John Newcomb.

233: poster for the Social Security Administration. Art director: Sheldon Cohen. Artist: Jerry Dadds/The Eucalyptus Tree Studio, Baltimore.

234: poster for a performance of Agatha Christie's play, "The Mousetrap." Agency: Sankowich/Golyn Productions,

Inc. Art director: Tom Tieche. Photographer: Larry Kunkel.

234: article, "Bone," for *Penthouse International* magazine. Art director: Joe Brooks. Illustrator: Alan Magee.

235: ecology poster for the state of Colorado. Art director: Warren Johnson. Photographer: Howard Sokol.

235: cover for an information booklet published by Ciba-Giegy, "Family in Distress." Art director: Ron Varetzis. Illustrator: Rick McCollum.

236: newspaper advertisement for the American Insurance Association, "... the shell game." Agency: Ketchum Advertising, Inc. Art director: Joseph H. Phair. Photographer: J. Barry O'Rourke.

236: experimental photo of shrouded figure by Laurence Sackman.

237: calendar showing hen on bottle of eggs. Photographer: Turin Studios, Inc.

237: cover showing knight with pen plume produced for *Graphis Annual 82/83*. Art director: Walter Herdeg. Illustrator: André François.

238: cover for *Graphis Posters 83*. Art director: Walter Herdeg. Illustrator: Seymour Chwast.

239: poster for a performance of the play, "Amadeus" at the Schiller Theatre in Berlin. Illustrator: Holger Matthies.

239: poster for an art studio, 20-20 Vision, Inc., in New York: "Ten happy summers." Art director: Theo Welti. Photographer and illustrator: Jacqueline Rose.

240: advertisement showing bridge above water for Champion Papers, Inc. Illustrator: Richard Hess.

240: poster for a production of "Anna Livia" at the Contemporary Theatre in Wroclaw, Poland. Illustrator: Jan Jaromir Aleksiun.

241: poster "A Scena Aperta" for *l'Unita*, the Italian Communist Party newspaper. Art director and illustrator: Lauro Giovanetti. Photographer: Alfredo Ferrari.

241: poster for a glass exhibition in Paris. Illustrator: Tomek Kawiak.

242: poster advertising Charvoz paint colors. Illustrator: Milton Glaser.

242: cover for *Graphis Annual 83/84*. Art director: Walter Herdeg. Illustrator: Jean-Michel Folon.

243: poster advertising the ninth annual Columbia University Holiday Crafts Fair. Agency: Chermayeff and Geismar, Inc. Illustrator: Ivan Chermayeff.

243: poster for the Theater der Welt in Cologne, Germany. Illustrator: Heinz Edelmann.

244–246: drawings by Bruce Bacon.

247: cover showing a television fan who becomes a hero. Produced for *Pushpin Graphic* magazine. Art director: Seymour Chwast. Illustrator: Elwood Smith.

LOW BUDGETS: CHAMPAGNE DESIGN WITH BEER MONEY

248: cover about cost containment for *Medical Laboratory*

Observer magazine. Art director: Tom Darnsteadt. Photographer: Walter Wick. Artist: Judith Jampel.

250: article about unfreezing third party (insurance) payments to pharmacists for *Drug Topics* magazine. Art director: Tom Darnsteadt. Photographer: Walter Wick.

251: article about a fraud that involved pension funds. Produced for *Medical Economics* magazine. Art director: Barbara Groenteman. Photographer: Stephen E. Munz. Artist: David Cayce.

252: article about year-around tax strategy for *Medical Economics* magazine. Art director: Barbara Groenteman. Illustrator: John Trotta.

252: cover about saving routine costs in the laboratory for *Medical Laboratory Observer* magazine. Art director: Roger Dowd. Photographer: Walter Wick.

253: advertisement for a financial service. Illustrator: Stanislaw Zagorski.

259: photo of John Newcomb by Charlotte Newcomb.

About the author

John Newcomb lives in Stamford, Connecticut, with his wife, Charlotte, and two children. He heads the art department for Medical Economics Company, a producer of medical magazines owned by International Thomson Organisation, Ltd.; the editorial office of Medical Economics is located in northern New Jersey. Newcomb was born in Topeka, Kansas, and obtained a BFA degree from Kansas University and a British National Design Diploma while studying in London on a Rotary Foundation Fellowship. He has worked as an editorial art director, promotion designer, and freelance illustrator in the New York metropolitan area for more than two decades. Recipient of numerous national design awards, he lectures on the subjects of art directing and creativity and is a member of the teaching faculty of the Folio *magazine* Face-to-Face *publishing seminar series held each fall in New York.* The Book of Graphic Problem-Solving *is his first book.*